Inorganic Massive Batteries

Energy Storage – Batteries, Supercapacitors Set

coordinated by
Patrice Simon and Jean-Marie Tarascon

Volume 4

Inorganic Massive Batteries

Virginie Viallet
Benoit Fleutot

iSTE

WILEY

First published 2018 in Great Britain and the United States by ISTE Ltd and John Wiley & Sons, Inc.

ISTE Ltd
27-37 St George's Road
London SW19 4EU
UK

www.iste.co.uk

John Wiley & Sons, Inc.
111 River Street
Hoboken, NJ 07030
USA

www.wiley.com

MIX
Paper from responsible sources
FSC www.fsc.org FSC® C013604

Library of Congress Control Number: 2018930592

British Library Cataloguing-in-Publication Data
A CIP record for this book is available from the British Library
ISBN 978-1-84821-724-9

Contents

Introduction

I.1. Energy issues

Originally dedicated to survival and dietary requirements, global energy consumption had been in equilibrium for over 12,000 years after the discovery of the first energy source in the form of fire. This balance was considerably affected by the industrial revolution of 1850. Since then, energy demands have been increasing steadily due to the exponential development of new technologies, better standards of living and growth of the world's population. All the energy consumed to satisfy our various needs comes from forms of the so-called primary energies, that are either non-renewable (fossil fuels such as coal, oil, natural gas, but also uranium) or renewable (hydro, wind, marine, geothermal and solar energies, including biomass). The evolution of the global consumption of these different sources is presented in Figure I.1 [IEO 16].

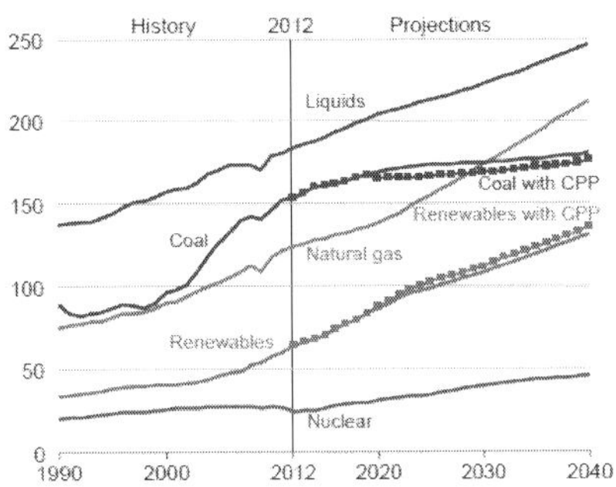

Figure I.1. *Global energy consumption between 1990 and 2040 (quadrillion Btu) by energy source [IEO 16]. The dotted lines show projections of the effects of the US Clean Energy Plan. For a color version of this figure, see www.iste.co.uk/viallet/batteries.zip*

Fossil fuels still account for this demand, and in 2014, 78.3% of the energy consumed was still from coal and oil and only 19.2% from alternative energy sources (Figure I.2) [REN 16]. Although extremely weak, these results are encouraging considering that in 1973, renewable energies accounted for only 0.1% of global energy consumption [WEO 12].

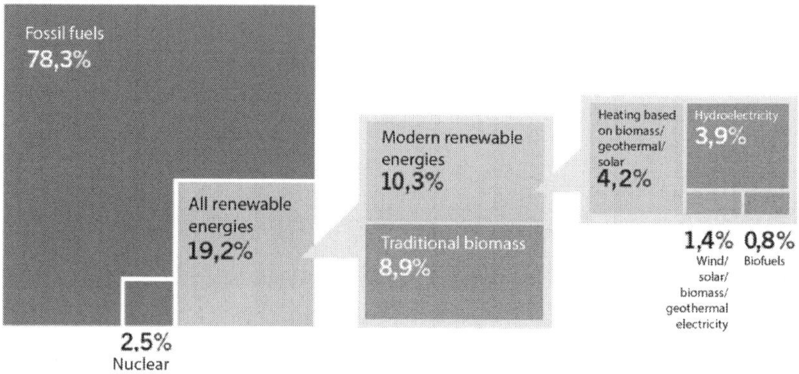

Figure I.2. *Estimated share of renewable energy in overall global energy consumption, 2014 [REN 16]. For a color version of this figure, see www.iste.co.uk/viallet/batteries.zip*

The replacement of these fossil resources is the major challenge of the 21st Century. The consumption of these resources has caught up with their production and their exhaustion is only a matter of decades away. Nevertheless, it appears that new sources of oil are possible, but at much higher costs than today. Moreover, even if fossil energy resources were not an issue, they should be limited because they are harmful to the environment mainly due to the gases emitted upon their combustion. It is imperative to turn to renewable energies.

The main disadvantage of renewable energies is that they are intermittent and that their production fluctuates during the day and according to the weather (i.e. for solar and wind in particular). They cannot power a power grid on demand. It is therefore necessary to develop new systems capable of reversibly storing the energy produced by these alternative resources. To fulfill this function, electrochemical cells appear to be the ideal storage system. The latter allow reversible storage of the electrical energy in the form of chemical energy and have the advantage of being adaptable with regard to their size.

The storage of electrical energy involves, on the one hand, stationary systems and, on the other hand, embedded systems. Stationary installations are dedicated sites, usually high capacity storage systems (> a few megawatt hours), medium or high power storage systems (from 100 kW to GW), which support the (continuous) power grids and production of renewable energies (wind and photovoltaics). On-board installations are small capacity storage devices integrated into a mobile system, in particular in rechargeable electric and hybrid vehicles, multimedia and equipment [LAR 15, POI 11].

I.2. Lithium Ion cells

As shown in Figure I.3 [TAR 01], lithium ion cells are the most efficient of electrochemical storage batteries, both in terms of volume and mass density, despite a higher price than their competitors, nickel-cadmium and nickel-metal hydride.

After a quick overview of the operating principle and of the characteristic features of a lithium battery, the means of increasing the energy density of these batteries will be discussed.

I.2.1. *Operating principle of a battery*

The operation of these batteries is based on an intercalation chemistry demonstrated in the 1970s, improved over the years with the use of lamellar oxides of transition metals proposed by Goodenough [MIZ 80].

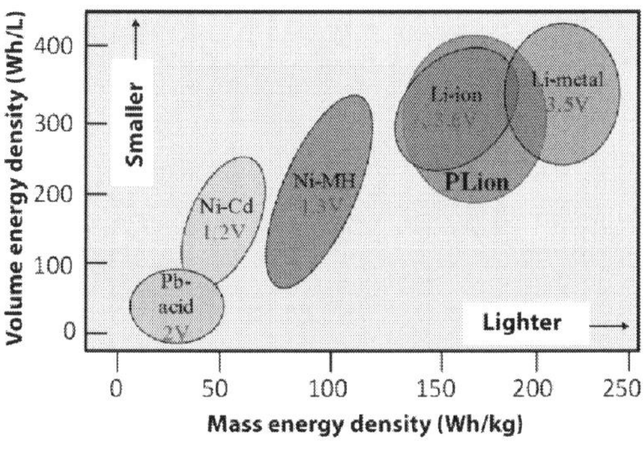

Figure I.3. *Comparison of mass and volume energy densities of the main electrochemical cell [REN 16]. For a color version of this figure, see www.iste.co.uk/viallet/batteries.zip*

Lithium metal has a very negative redox potential (–3.03V compared to ENH), combined with a high specific capacity (3862 mAh/g). Lithium is one of the lightest elements (M = 6.94 g/mol) making it a negative electrode material which can be used in batteries with very high energy densities. However, the widespread commercialization of rechargeable lithium batteries is hampered by its limited cyclability and inherent safety issues. Like many metal electrodes, lithium is morphologically dynamic. Its surface morphology changes, since during electrochemical cycling a part of the lithium migrates to the positive electrode to react and is then repositioned on its surface in a heterogeneous manner leading to the growth of a porous foam, a change in volume and sometimes dendrites. Lithium's very negative potential leads to the reduction of the electrolytes and the formation of an interface called the SEI, the creation of a new surface which causes the progressive deterioration of the electrolyte through redox reactions. This can incude secondary reactions that can cause safety concerns [DOL 02]. As a result, it has rapidly been replaced by graphite carbon: new Li-ion batteries, where lithium is present in an ionic and non-metallic form, have seen their performance rapidly approach those of lithium batteries and were marketed for the first time in 1991 by Sony [NAG 90]. These $LiCoO_2$/graphite batteries still constitute the vast majority of Li-ion batteries sold on the market.

The working principle of a Li-ion cell, with $LiCoO_2$ as the positive active material, carbon graphite as negative active material and LP30 ($LiPF_6$ 1M, EC:DMC 1:1) as a liquid electrolyte for example, is presented in Figure I.4 [TAR 15]. A separator, not shown in the diagram, is placed between the two electrodes in order to avoid direct contact and to allow the transfer of the Li^+ ions. This type of cell is typically used in Li ion cell phone batteries for example. During charging, at the positive electrode, the oxidation reaction releases Li^+ ions within the electrolyte which migrate and intersperse between the graphite sheets (negative electrode) thanks to the addition of electrons by the external circuit. During discharge, the disinsertion of lithium is accompanied by the release of an electron in the external circuit which supplies electricity to the device attached to the battery. The negative electrode is fixed to a copper sheet, which improves the electric conductivity of the battery [BRU 97]. The lithium ions move reversibly between the electrodes during cyclability; this battery is often known as a "rocking-chair battery".

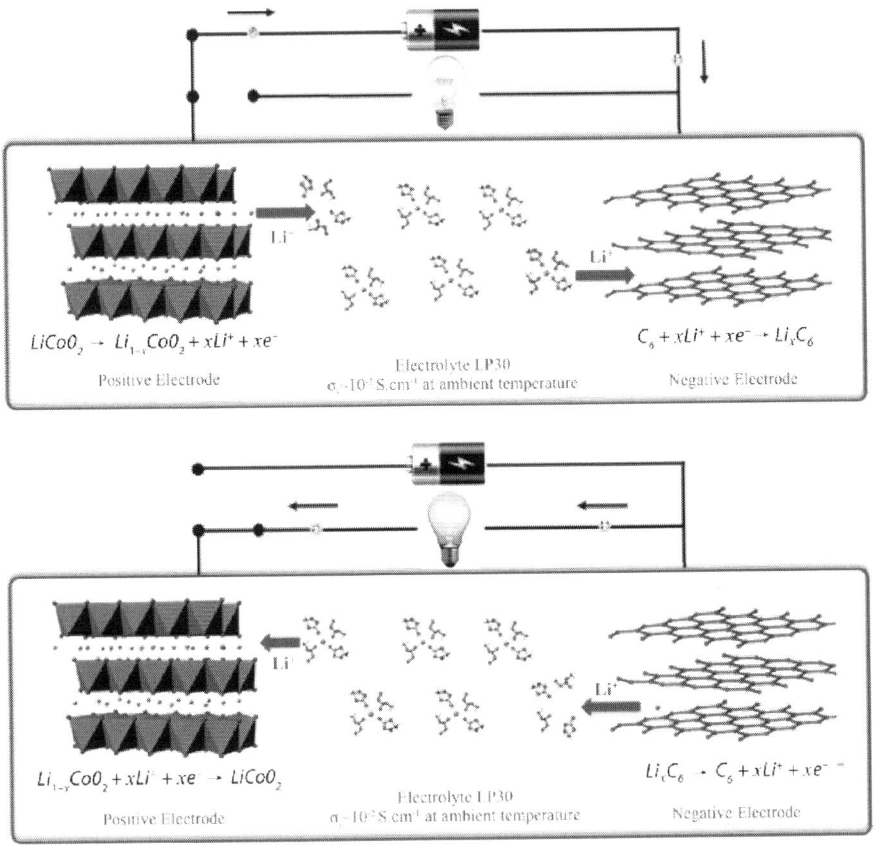

Figure I.4. *Working principle of a Li-ion battery during charging (top) and discharge (bottom) [TAR 15]. For a color version of this figure, see www.iste.co.uk/viallet/batteries.zip*

I.2.2. *Main characteristics of a battery*

The main characteristics of a battery are as follows:

– capacity;

– operating voltage;

– stored energy and energy density;

– current density and the maximum current applicable;

– cyclability;

– calendar life-time;

– self-discharge;

– temperature resistance.

I.2.2.1. *Capacity*

The capacity Q (in Ah) is the amount of electricity or electrical charge supplied or stored by the battery. It is the product of the intensity I of the current which passes through the generator (in amperes A) per time of passage of the current t. If t is in seconds, Q is in Coulombs (C) and if t is in hours, Q is in ampere-hours (Ah) (1 Ah = 3600 C) :

$$Q = I \times t.$$

The capacity can be expressed as a function of the number of moles of electrons inserted or deinserted per mole of active material:

$$Q = \frac{m \cdot F \cdot x}{3600 \cdot M}$$

where m : mass of active material (g),

F : Faraday number = 96500 $C.mol^{-1}$,

x : number of moles of electrons exchanged per mole of active material,

M : molar mass of the active material ($g.mol^{-1}$).

This number of exchanged electrons x corresponds to the number of Li^+ ions introduced into the host material according to the reaction, with M_t being the transition metal involved:

$$M_t + xLi^+ + xe^- \underset{\text{charge}}{\overset{\text{discharge}}{\rightleftarrows}} Li_xM_t$$

We also define the surface ($\mu Ah/cm^2$) or volumic ($\mu Ah/cm^2/\mu m^1$) capacity as an important criterion in the case of micro-batteries and "all-solid" batteries.

I.2.2.2. *Voltage*

The theoretical voltage E_{OC}, or open circuit voltage (OCV), of a battery depends on the thermodynamic properties of the electrode materials. It is defined by the equation:

$$E_{OC} = \varphi_c - \varphi_a$$

where φ_c : standard voltage of the positive electrode,

φ_a : standard voltage of the negative electrode.

When connected to an external circuit, the operating potential is always lower than the E_{OC} value due to an ohmic drop and polarization induced at the electrodes.

The operating potential (or voltage) U of a battery is mainly determined by the choice of combination of the positive and negative electrode materials. The evolution of this potential during the discharge and the charging of a lithium battery is presented in Figure I.5 [LEV 09].

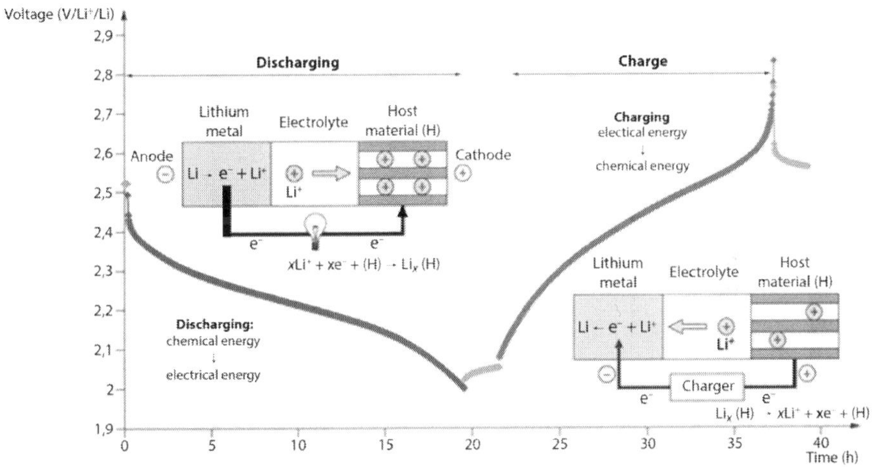

Figure I.5. *Evolution of the voltage of a lithium battery during its operation [LEV 09]. For a color version of this figure, see www.iste.co.uk/viallet/batteries.zip*

I.2.2.3. *Power, energy and energy density*

The power consumed P (in W) is equal to the product of the voltage U (in V) of the battery by the current I (in A) which it delivers:

$$P = U \times I.$$

The energy E stored in the battery is equal to its electrical charge Q multiplied by the mean voltage U under which this charge is discharged, but also to the product of the power P (in W) absorbed by the operating time t (in s or h) :

$$E = Q \times U = P \times t$$

The energy stored is usually measured in watt-hours (Wh) but the official unit (SI) is the joule (t expressed in s).

The amount of energy discharged from a battery is called the discharge depth. It is given as a percentage of its capacity.

The peak power density, or specific power, corresponds to the maximum power related to the mass of the battery, and is expressed in watts per kilo (W/kg). The mass density, or specific energy, corresponds to the amount of energy (Wh/kg) that the cell can restore per unit mass. The volumetric density, or volumetric energy density, corresponds to the amount of energy (Wh/m^3) that it can restore relative to its volume. Wh/dm^3 or Wh/L are more frequently used.

I.2.2.4. *Current density and maximum current charge*

The current applied to a battery is measured in amperes A but for micro-batteries and "all-solid-state" batteries, the current density is expressed, i.e. in $\mu A/cm^2$.

The C rate expressed as C/n is also used, where n corresponds to the number of hours required to complete a discharge or a full charge. For example, if the battery is discharged for 1 hr, the current density is 1C. If 2 hrs of discharge is required, the current density is C/2.

I.2.2.5. *Cyclability (lifetime)*

Cyclability, expressed as the number of charge/discharge cycles, characterizes battery life and the reversibility of internal processes.

The loss of capacity during cyclability is linked to several phenomena. Structural or morphological modifications of the electrode materials, mechanical damage due to material transfer or permanent and cyclical evolution generally due to chemical reactions between the various internal components may be the cause of this loss of capacity.

Many parameters influence the cyclability of a battery such as the electrode materials used, the electrolyte, the discharge depth per cycle, the current density and the operating (or storage) temperature.

For "all-solid" thin-film batteries, excellent cyclabilities have been reported with over 10,000 cycles performed [LEV 09].

I.2.2.6. *Self-discharge*

A battery will discharge slowly even when not in use. The rate of self-discharge increases rapidly with temperature and when the battery ages. The origin of self-discharge arises from numerous secondary reactions occurring at the electrodes.

I.2.3. *Increasing the energy density*

Current standard 18650 cylindrical Li-ion batteries have volumetric and gravimetric energy densities of up to 770 Wh/L and 260 Wh/kg, respectively. The standardized designation of a Li-ion battery is a five-digit number, the first four of which define the dimensions of the battery and the last one its shape. In the case of an 18650, the final 0 indicates a cylindrical shaped battery. The first two digits give the diameter 18 mm and the next two give the length 65 mm (with an approximate volume of 16 cm^{-3} and a mass of 48 g).

As energy demands increase, more efficient Li-ion batteries are required, which must deliver even higher energy densities. The energy density of a battery being the product of its intrinsic capacity and its average potential, there are two possible ways to increase its energy density:

– the first is to develop new electrode materials with a high intrinsic capacity. As shown in the theoretical study carried out by the Ceder's team and the results of which are presented in Figure I.6, numerous electrode materials could achieve extremely high energy densities [CED 11].

Unfortunately, this same study shows that high capacity tends to lead to instability of the materials and thus to their degradation;

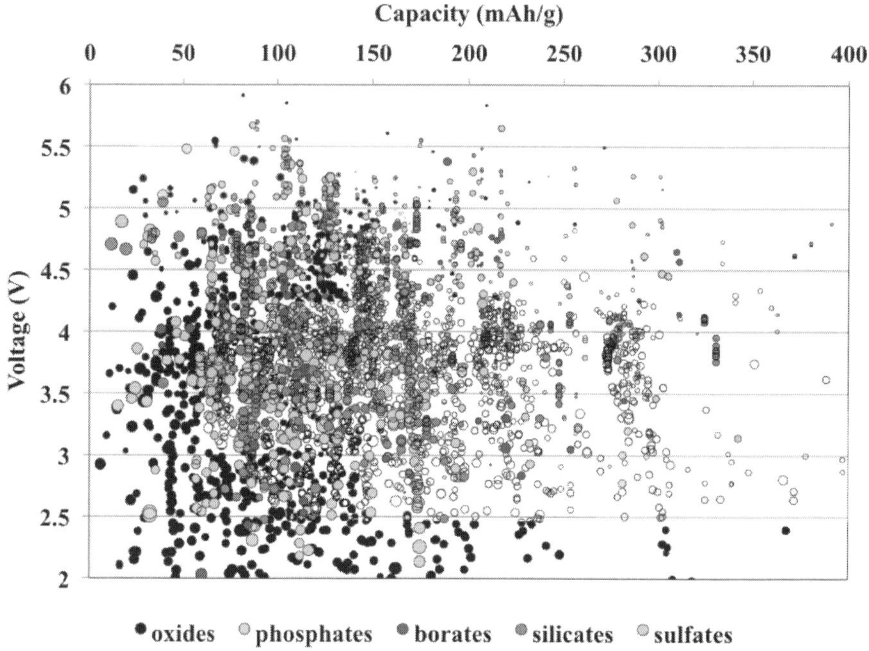

Figure I.6. *Overview of the electrochemical properties of materials of positive electrodes: past, present and future [CED 11]. For a color version of this figure, see www.iste.co.uk/viallet/batteries.zip*

– the second way would be to increase the potential difference between the two electrodes. Since graphite has the lowest voltage to date, besides lithium metal, it is therefore necessary to increase the redox potential of the positive electrode. For example, $LiCoPO_4$ has a potential of 4.9 V for a capacity of 169 mAh/g, making this material an excellent choice to replace $LiCoO_2$. Unfortunately, the organic liquid electrolytes used to transport lithium within the battery are instable at potentials greater than 4.2–4.4 V *versus* Li^+/Li. Use beyond this potential range can result in battery explosion and release of toxic gases (e.g. HF). As a result, it appears that to date the energy density of rocking-chair batteries seems to have reached a maximum limit as shown in Figure I.7 [JAN 16].

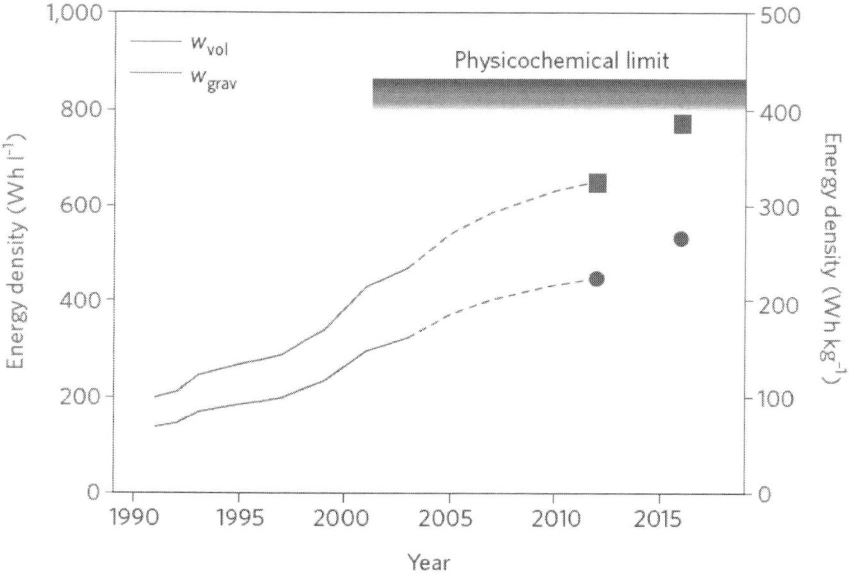

Figure I.7. *Evolution of the energy densities of a standard 18650 battery over the last 25 years [JAN 16]. For a color version of this figure, see www.iste.co.uk/viallet/batteries.zip*

Numerous studies are being carried out to replace this technology with Li–S or Li–air technology which, theoretically, perform much better than current Li-ion batteries [BRU 12].

A possible solution to increase the energy density of Li-ion batteries would be to replace the liquid organic electrolytes with inorganic, non-flammable materials with a much higher potential stability range. However, the use of Li-ion batteries in large-scale devices is restricted by safety problems related to the nature of the liquid electrolytes. On the other hand, lithium is not very abundant in the Earth's crust and ore deposits are unevenly distributed across the planet, which makes this element a resource with high economic or even geopolitical stakes, like oil.

Resources are estimated at 40 million tons with about 9 million in Argentina and Bolivia, 7.5 million in China and Chile, and 2 million in Australia and Canada [MCS 17].

I.3. Importance of all-solid-state batteries

A battery is an assembly of cells in series and/or in parallel each consisting of a positive electrode, a negative electrode and an electrolyte with a separator. Depending on the nature of this electrolyte, Li-batteries can be divided into three categories:

– conventional (liquid electrolyte) batteries;

– polymer batteries (with a viscous polymer electrolyte depending on the temperature, hence sometimes incorrectly classified as "all-solid-state" battery);

– "all-solid-state" batteries (with solid electrodes and electrolyte).

The comparative diagrams of a conventional lithium-ion battery and an all-solid-state battery are presented in Figure I.8 [JAN 16]. The conventional battery shown in the center and denoted LIB consists of a porous negative electrode, typically graphite (gray spheres) and a positive electrode, typically a transition metal oxide (violet spheres), 100 μm thick between which is placed a separator approximately 10 μm thick (gray band). The liquid electrolyte impregnates the electrodes and the separator, ensuring the rapid transfer of ions. In an all-solid-state Li-ion battery, represented on the right and denoted Li-SSB, the liquid electrolyte is completely replaced by a solid electrolyte (*orange spheres) within the electrodes and the separator impregnated with the liquid electrolyte is replaced too. The diagram on the left represents an all-solid-state battery with lithium metal as the negative electrode, with which it is possible to significantly increase the energy density.

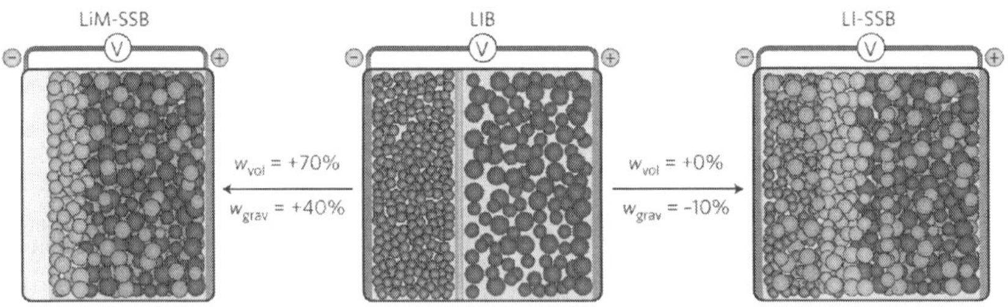

Figure I.8. *Comparative diagrams of a conventional lithium-ion battery (LIB in the center) and an all-solid-state battery with lithium metal (LiM-SSB on the left) or a composite negative electrode (Li-SSB on the right). w_{vol} and w_{grav} are the volumetric and gravimetric energy densities, respectively [JAN 16]. For a color version of this figure, see www.iste.co.uk/viallet/batteries.zip.*

The use of a solid electrolyte in an electrochemical cell is not a new idea. The first studies date back to the beginning of the 20th Century with the work of Tubandt and Lorenz on the α-AgI conductor [TUB 14]. Much more research has been done on copper conductors, but the electrode materials used have too low energy densities making the electrolytes uninteresting for commercial applications.

The interest in solid electrolytes was revived in the 1960s thanks to the discovery of the sodium conductor, beta-alumina [YAO 67]. The discovery of this material has allowed the development of the first "solid-state" batteries based on molten sodium: Na–S and Zero Emissions Batteries Research Activity (ZEBRA) batteries. These batteries have energy densities between 140 and 240 Wh/kg and use cheap, abundant materials. Almost 90% of the total mass of the battery is recyclable. In addition, the use of a solid electrolyte allows zero self-discharge of the cell because the latter is an electronic insulator. The main issue with these cells is that they must be used between 270 and 350 °C, inducing an on-board heating system that greatly reduces the actual energy density of these batteries (taking into account the total mass of the system). These operating temperatures are nevertheless necessary in order to obtain sufficient ionic conduction of beta-alumina and allow for the use of molten sodium and sulfur. However, over 200 installations of Na–S batteries have been implemented around the world for stationary applications and several hundred commercial vehicles, such as the Citroën Berlingo first, have been equipped with ZEBRA batteries sold by the Swiss company MES-DEA [AVE 10].

Li ion batteries have some intrinsic limitations, which could be solved by using solid electrolytes:

– Safety issues: Liquid electrolytes, such as LP30 (LiPF6 1M EC:DMC 1:1) for example, can decompose in contact with air, moisture or when the temperature exceeds 60 °C. The gases resulting from this decomposition can be toxic (release of hydrofluoric acid in air [ROT 12]) and catch fire, especially in the event of a short circuit or surcharge. There are several examples of fire caused by Li ion batteries. In February 2013, the Boeing 787 Dreamliner aircrafts were banned from flying worldwide for several weeks after several fires involving Li-ion batteries with liquid electrolytes. Drastic safety measures must therefore be observed to prevent any risk of leakage.

– Organic liquid electrolytes do not allow applications at high temperatures because of their boiling temperatures of between 90 and 250 °C depending on the compositions. High temperature applications, such as in petroleum boreholes, are limited by the use of a liquid electrolyte.

– Liquid electrolytes do not allow Li metal to be used as a negative electrode, resulting in loss of potential.

– The presence of a liquid electrolyte, even if simply impregnated in a separator, can lead to stresses on the conditioning and assembly of the cells, which could be simplified by the use of solid electrolytes. This argument is particularly evoked by Toyota (Figure I.9) [TOY 95-18], or by the National Institute of Advanced Industrial Science and Technology (AIST) [AIS 10] (Figure I.10).

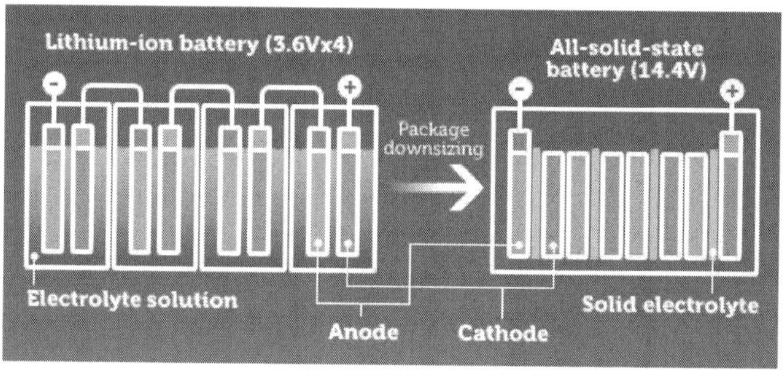

Figure I.9. *Schematic representation of the volume gain brought about by using "all-solid-state" technology compared to conventional liquid electrolyte technology [TOY 95-18]. For a color version of this figure, see www.iste.co.uk/viallet/batteries.zip*

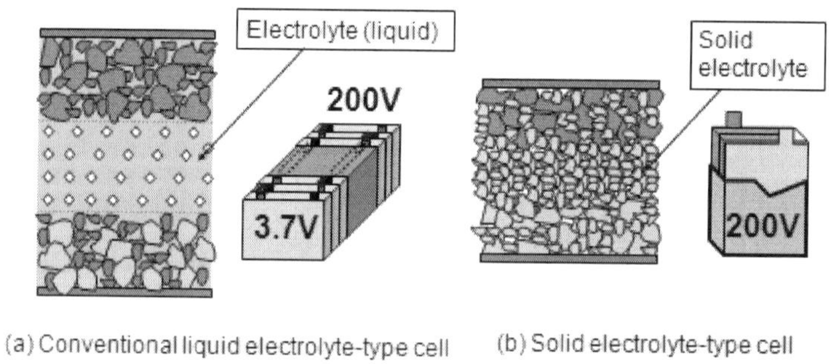

Figure I.10. *(a) Comparison of a conventional cell with a liquid electrolyte and (b) an all-solid-state cell [AIS 10]. For a color version of this figure, see www.iste.co.uk/viallet/batteries.zip*

I.4. Challenges of all-solid batteries

Two major technological challenges, common to any type of "all-solid" design, hinder the development of such batteries: the low intrinsic ionic conductivity of solid electrolytes and the difficulty of obtaining good solid/solid interfaces within the composite electrodes and the complete system [LAL 15].

Ionic conductivity, both within electrode materials and the electrolyte, is a key point in the operation of a battery. It describes the mobility of ions throughout the system and greatly influences performance, especially the power of the system. This conductivity depends on the conductivities in the three layers of the battery, which are themselves dependent on the intrinsic conductivities of the materials and at the grain boundaries (which can ultimately be related to the solid/solid interfaces). However, the intrinsic conductivity of conductive lithium or sodium solids at room temperature is generally much lower than that of liquid electrolytes (10^{-2} S/cm). This is especially true for electrode materials. They are preferably restricted to the use of electrodes in the form of thin layers or composite electrodes. Overcoming this obstacle is thus dependent on further research into ionic conductors.

From a purely theoretical point of view, the solid/solid interfaces generated in "all-solid" batteries are preferable to solid/liquid interfaces generated in conventional batteries. Indeed, the activation energy of the charge transfer at the interface is lower for a solid/solid interface since it does not include a desolvation step of the Li^+ ion. However, in reality, these interfaces are not ideal and their qualities depend on the materials selected and the shaping technique used. As with conventional batteries, it has been shown that an incompatibility between the electrode material and the solid electrolyte material can lead to the formation of a fairly stable and resistive interface (SEI). On the other hand, even without reactivity, poor interfaces between grains (from the same material or not) increase the resistance to grain boundaries, which decreases the total conductivity of the battery. Overcoming this very complex obstacle is a major issue.

The main challenges posed by the development of all-solid technology are presented in Figure I.11 [JAN 16]. In this figure, lithium metal is used as a negative electrode, its use being able to significantly increase the energy density of such a battery thanks to its high specific capacity. However, a resistive interface (SEI) is formed between lithium and the solid electrolyte (SE), the resistance of this interface is represented by $R_{SE/SEI}$ (left circle). In this same circle, R_{SE} and $R_{SE,gb}$ represent the resistive contribution of the electrolyte grains and grain boundaries. During the charging/discharging cycles, the lithium metal is deposited in an inhomogeneous manner, resulting in the formation of dendrites which can cause a short circuit (bottom left circle).

Most electrolytes react with positive electrode materials and must be protected with a coating. This coating is also applied to avoid the formation of a lithium-depleted zone on the surface of the electrolyte (Li^+-depleted SE).

The blue arrows indicate the mechanical pressure required to avoid loss of grain contact due to the change in material volume during lithiation/delithiation cycles.

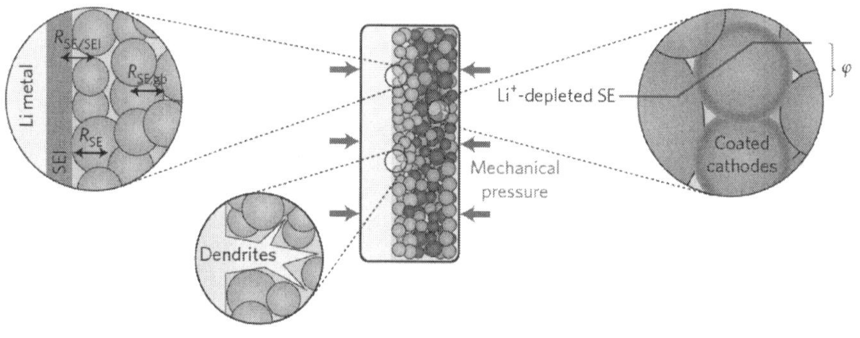

Figure I.11. *Main challenges in the development of all-solid-state battery technology [JAN 16]. For a color version of this figure, see www.iste.co.uk/viallet/batteries.zip*

I.5. "Thin film" batteries and 2D micro-batteries

The history of thin-film batteries in the 2D configuration began in 1982 when the Japanese company Hitachi announced a "Thin-Film All-Solid-State Battery". This 4 × 4 mm battery consisted of a layer of CVD-deposited TiS_2 (1–3.7 μm), a layer of solid LISICON ($Li_{3,6}Si_{0,6}P_{0,4}O_4$) electrolyte deposited by cathode sputtering (2–4 μm) and a lithium layer obtained by vacuum evaporation (2–4 μm). This assembly delivered a voltage of 2.5 V for a current density between 3 and 16 μA/cm² [KAN 83]. Figure I.12 shows the simplified diagram of a micro-battery in thin layers, prepared by successive depositions of different layers of thickness varying from a few nanometers to a few micrometers. Full stacking includes current collectors, active layers and encapsulation and does not exceed a few tens of micrometers [LEV 09].

This technology was quickly taken over by American groups such as Ever-Ready and Bellcore, who replaced LISICON with chalcogenide glass such as Li_3PO_4–P_2S_5 or $Li_4P_2S_7$ [AKR 86, AKR 88]. In France, the teams of Ribes and Levasseur, who had demonstrated their expertise in the synthesis of conducting glasses and glass-ceramics, used their experience to assemble micro-batteries [CRE 89, MEU 89]. Levasseur's work has enabled the *Laboratoire de Chimie de Bordeaux* (now ICMCB) to become a reference in the field of 2D micro-batteries.

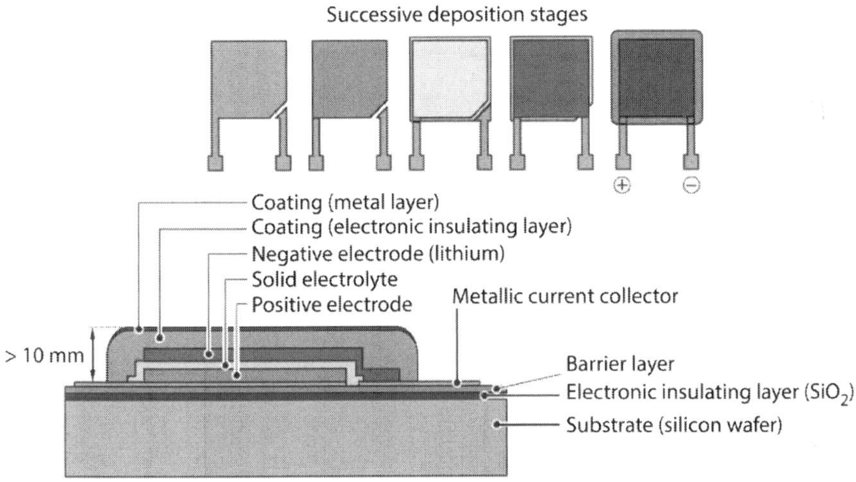

Figure I.12. *A micro-battery [LEV 09]. For a color version of this figure, see www.iste.co.uk/viallet/batteries.zip*

The discovery of LiPON by the teams at the Oak Ridge National Laboratory in the United States has revived the dynamics of this technology and allowed for a considerable increase in performance. LiPON (or $Li_{3-x}PO_{4-x}N_x$) is a glass with a conductivity of 2×10^{-6} S/cm at 298 K and an electrochemical stability up to 5.4 V which can easily be obtained in the form of thin films by sputtering. The use of this electrolyte makes it possible to increase the number of positive/negative electrode combinations which, in turn, allows the voltage to be increased (from 2 to 5 V), as well as the current density to 1 mA/cm^2 and the capacity retention (30,000 cycles reported for $LiCoO_2$) of thin film batteries [BAT 95, FLE 11a, WAN 96].

Research by Fleutot *et al.* into the effect of Solder-Reflow (heat treatment during the assembly of printed circuits) on the conduction properties of LiPON [FLE 11b, FLE 12] have shown an improvement in the ionic conduction properties, which is also accompanied by an increase in the activation energy by a modification of the chemical bonds within this vitreous electrolyte. A new vitreous electrolyte was proposed by the same team in order to overcome this degradation [FLE 13].

Numerous studies have been carried out to replace lithium at the anode, its use giving rise to safety issues, due to the risk of melting during the assembly of the printed circuits (Solder-Reflow process) of the electronic components of the battery [BAB 99, NEU 00, NEU 99].

In the literature, it seems that micro-battery technology allows for a substantial increase in performance of certain electrode materials, in comparison with conventional batteries using a liquid electrolyte. This is the case, for example, for silicon. Indeed, Pecquenard *et al.* proposed a Li/LiPONB/Si battery in 2012, with a remarkable capacity retention over 1,500 cycles, combined with a Coulombic efficiency close to 100% [PHA 12]. These excellent electrochemical properties, compared to those of conventional batteries using an organic liquid electrolyte, are partly induced by the "all-solid" technology. The use of a vitreous solid electrolyte makes it possible to limit the reactions at the electrode interface (i.e. SEI). The strong adhesion of silicon with LiPONB prevents the formation of cracks due to variations in volume.

Paradoxically, the size of micro-batteries, which is their greatest quality, is also their biggest problem. Indeed, as reported by Dudney in 2005 [DUD 05], the energy and power densities of micro-batteries are limited by the thickness and the composition of the positive electrode, as illustrated in Figure I.13. Due to the wide diffusion of the lithium in its sheets, $LiCoO_2$ allows for the assembly of the micro-batteries with the highest power and energy densities (1 mW/cm^2 per 1 mWh/cm^2).

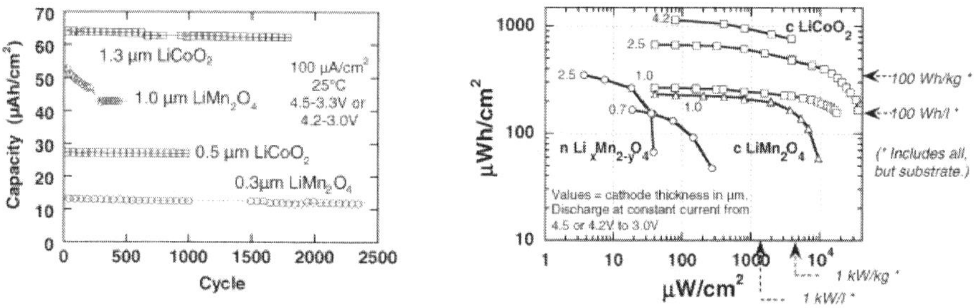

(a) (b)

Figure I.13. *Influence of the cathode material and electrode thickness on the performance of a micro-battery: capacity retention (a) and power and energy density (b) [DUD 05]*

Increasing the thickness of electrodes and electrolyte and obtaining massive all-solid batteries is highly recommended for use in large-scale storage systems. However, increasing the thickness of the battery requires the electrolyte to have high ionic conduction and a very low interfacial resistance between the electrolyte and the electrode.

Currently, 2D micro-batteries are capable of delivering surface capacities of approximately 100–200 µAh/cm^2. To meet the goals of "all-solid" state technology, it is necessary to switch from a surface capacity to a volume capacity. The thin-film deposition techniques adopted for the manufacture of 2D micro-batteries are unfit for this transition because, in addition to the cost of manufacture, the deposition of an overly thick layer leads to delamination and the appearance of cracks in the electrode. It is therefore necessary to consider other solutions for the manufacture of 3D micro-batteries and more massive batteries. The latter, which are of macroscopic dimensions, are called monolithic batteries.

1

Anatomy of an All-Solid-State Battery

The principle of an all-solid battery is very similar to that of a conventional battery. It comprises a negative and a positive electrode, and an electrolyte, the latter being solid [BRU 97, YOO 13]. The presence of a solid electrolyte causes changes in the synthesis of various constituent elements, shaping and stresses with respect to a conventional battery.

In a conventional battery, the liquid electrolyte is in close contact with the components of the positive and negative electrodes, ensuring a good transfer of the Li^+ ions: the percolation of the Li^+ ions is therefore optimal with a liquid. In fact, the electrodes bathe in this electrolyte. In the case of an all-solid-state battery, this percolation must also be ensured and the solid electrolyte must therefore be integrated within the electrodes, just as the liquid electrolyte penetrates the latter in the case of conventional batteries. In conventional Li-ion batteries, the electrode can be regarded as a composite, that is to say, a mixture of active material, for example, $LiCoO_2$, and a conductive additive, such as carbon. In the case of an all-solid-state battery, the electrodes must contain an electrolyte. These are then referred to as composite electrodes which therefore comprise three constituents, as shown in Figure 1.1.

The proportion of various constituents, the particle size of the constituents together and their mixing will have an impact on the performance of the composite, and therefore, on the overall cell performance.

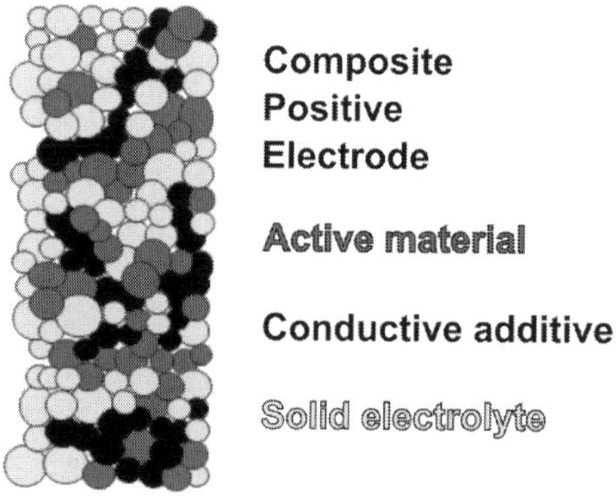

Figure 1.1. *Composite electrode comprising an active material (in purple), a conductive additive (in black) and a solid electrolyte (in light blue). For a color version of this figure, see www.iste.co.uk/viallet/batteries.zip*

All-solid-state batteries have two main morphologies as shown in Figure 1.2. The difference between the two is essentially at the negative electrode, which can be composite (left) or metallic (Li, In, Li–In alloy, etc.) (right) according to the nature (compatibility) of the solid electrolyte (with respect to lithium). The solid electrolyte thus transfers the Li^+ ions but also acts as a separator between the composite electrodes.

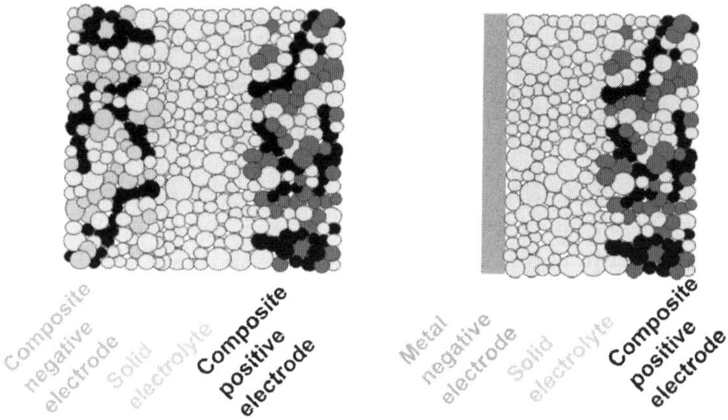

Figure 1.2. *All-solid battery resting on a composite positive electrode, a solid electrolyte and a composite negative electrode (left) and a metal electrode (Li, In, Li–In alloy, etc.) (right). For a color version of this figure, see www. iste.co.uk/viallet/batteries.zip*

Two important parameters to consider when replacing a liquid electrolyte with a solid electrolyte are the following:

– the first is the appropriate choice of solid electrolyte. The latter must have the highest possible ionic conductivity and a wide electrochemical stability window;

– the second is the critical management of the interfaces between the solid electrolyte and the active material within the composite electrode. It will be necessary to consider two types of interfaces: between the composite electrode and the solid electrolyte on the one hand and within a composite electrode on the other.

We are interested here in massive all-solid-state batteries for which the thickness of the electrolyte layer is approximately a few hundred micrometers, unlike in micro-batteries for which this layer is between a hundred nanometers and a few microns thick.

1.1. Constituents of an all-solid battery

1.1.1. *Nature of solid electrolytes: required qualities*

Since the performances of all-solid batteries depend in particular on the diffusion of the ions within the electrolyte, it must have a very good ionic conductivity, a very low electronic conductivity and a high chemical and electrochemical stability [THA 06b]. The choice of electrolyte is therefore based on the following criteria:

1) chemical and thermal stability;

2) ion transport number;

3) strength and mechanical stability.

1) *Chemical and thermal stability*

Since the solid electrolyte is placed between the electrodes, it is essential that no reaction detrimental to performance occurs at the interfaces of the composite electrodes with the electrolyte [AGR 08]. The electrolytes must be capable of operating perfectly in a voltage range as wide as possible, at least between 0 V and approximately 4–5 V [AGR 08, PAR 13]. Therefore, chemical and electrochemical stabilities are essential properties.

Usually, rechargeable lithium batteries used for military purposes are subjected to temperatures between –50 and 80 °C, and the electrolyte must be capable of operating within this temperature range without showing any thermal degradation [AGR 08]. Thermal stability is therefore another important factor that must be taken into account when selecting the electrolyte.

2) *Ion transport number*

The transference number, or ion transport number, indicates the contribution of different ions to the total electric current transported by the electrolyte [AGR 08, EVA 87], or the number of transported ions relative to the total number of ions. This number must be close to 1 or ideally equal to 1 [AGR 08, QUA 11]. Indeed, the main purpose of the electrolyte is to allow the flow of ions (Li^+ cations in the case of Li-ion batteries) and to prevent electron transfer. In reality, the cation transfer number is usually close to 0.5, indicating that some of the necessary ionic species move through the electrolyte [AGR 08, BRU 92]. Therefore, the chosen electrolyte should have an ion transfer number as close as possible to 1, in order to lead to efficient cation transport, reduce polarization, and therefore, lead to higher power densities [AGR 08]. For example, a solid polymer lithium salt such as $(PEO)_{20}LiClO_4$ has a lithium ion transport number equal to 0.31, whereas for the electrolyte PEO-$LiAsF_6$ it is close to 0.44 [KAT 04]. Some polymer-gel films showed cation transport numbers greater than 0.95 [ZAI 13].

3) *Mechanical resistance*

The solid electrolyte must support a large mechanical load because it is the backbone of a solid battery. Stresses can develop due to a thermal cycle due to volume expansion/contraction of the active materials present within the composite electrodes during the charging/discharging processes but also during shaping, as well as containing possible dendritic growths. This is why the mechanical properties of solid electrolytes must be taken into consideration.

1.1.2. *Positive electrode materials*

The positive electrode material is important because, depending on the configuration of the system, it provides the battery with the ions necessary for the charging and discharging processes. The positive electrode material must be structurally stable during this process. The ionic conductivity of this material must also be good, as the charging and discharging processes involve the transfer of ions through the electrode [WHI 04, FER 10, GAM 71]. Lithium and cobalt oxide (LCO), with a stoichiometry $LiCoO_2$, is widely used in Li-ion batteries available on the market (Sony, Sanyo, Toshiba, etc.). It is also found in all-solid-state devices. LCO has a lamellar structure [JOH 58, AMA 96, ROS 93] adapted for the delithiation/lithiation process, with a potential of 3.9 V (vs. Li) and a theoretical capacity of 135 mAh/g corresponding to the reaction below:

$$LiCoO_2 \Leftrightarrow Li_{0.5}CoO_2 + 0.5\ Li^+ + 0.5\ e^- \qquad [1.1]$$

making it the material of choice [WHI 04, FER 10, MIZ 80]. However, it is relatively expensive to manufacture, in particular because of cobalt [ROS 13].

Lithium and manganese oxide $LiMn_2O_4$ is another compound used as a positive electrode active material in all-solid batteries [BRU 46, KAT 04]. Its spinel structure favors the migration of lithium ions during the lithiation and delithiation processes [WHI 04]. The specific capacity is 148 mAh/g with a potential of 3.8 V (vs. Li), capacity corresponding to the following reaction:

$$LiMn_2O_4 \Leftrightarrow Li_{1-z}\ Mn_2O_4 + z\ Li^+ + z\ e^- \qquad [1.2]$$

$LiMn_2O_4$, however, has some disadvantages, including a phase change during the ion transfer process.

$LiFePO_4$, a lithium-based phosphate, has the advantage of being less hazardous and less expensive to produce than other lithium oxide materials [WHI 04, YAN 03]. Its olivine structure (unidimensional chain of lithium ions) greatly favors ion transfer [WHI 04, FER 10]. On the other hand, the self-discharge is high, which reduces the lifetime of the electrode. The theoretical capacity of this iron phosphate is 170 mAh/g with an operating potential of the biphasic system $LiFePO_4/FePO_4$ equal to 3.5 V versus Li^+/Li.

1.1.3. *Negative electrode materials*

Lithium metal with its high theoretical capacity of 3840 mAh/g, its light weight and ease of ionization of lithium during charging is an extremely promising material as a negative electrode for rechargeable batteries. Unfortunately, in liquid electrolyte, lithium metal is deposited in the form of foam and not uniformly on the surface of the metal electrode. The application of a strong current can induce the formation of dendrites which pass through the electrolyte, and thus, cause short circuits in the cell, leading to the loss of the battery component. These irregular deposits can also break up by creating isolated microaggregates, leading to a loss of capacitance, since part of the lithium is no longer involved in the electrochemical processes and there are problems in terms of safety; lithium in the liquid electrolyte is highly reactive. All-solid-state batteries and the expected lack of dendritic growth in these systems revive interest in lithium metal. Although lithium is the ideal candidate, we will see that its use is limited to low densities because of the difficulty in establishing an interface that favors its dissolution and its redeposition. Indium, despite the excellent results reported in the literature, is not a likely choice for industrial applications because of its expensive price of 547 €/kg (on July 2, 2013, source: http://bourse.lesechos.fr).

Lithium titanate (LTO) with Li4T5O12 stoichiometry has been widely used in all-solid-state systems. Its spinel structure makes it easy to integrate lithium ions during the lithiation process [WU 03, GEN 09, AHN 05]. An advantage of its use resides in the lack of volume variations during the lithiation/delithiation processes. Its theoretical capacity is 175 mAh/g according to the following reaction:

$$Li_4Ti_5O_{12} + 3\,Li^+ + 3\,e^- \rightleftharpoons Li_7Ti_5O_{12} \qquad [1.3]$$

This reactivity follows a biphasic process which plateaus at 1.55 V. Moreover, despite the three Li^+ ions introduced into its structure, this compound exhibits a zero volume variation, hence its interest on Li-ion batteries.

Carbon, in particular graphite, with its lamellar structure favorable for the intercalation of Li^+ ions, and carbon-based materials are commonly used as anode materials in "all-solid" batteries [AZU 99, KAS 07, KOT 12]. Graphite can also easily be doped with other materials to improve its capacity which is 350 mAh/g. Soft carbons have also been integrated: their disordered structure favors the incorporation of Li ions [WU 03, AZU 99]. Graphitic carbons, in particular, are commonly used at the negative electrode of commercial lithium batteries by virtue of their higher specific charge and the more negative oxidation-reduction potential than those of metallic or polymeric oxides. They perform better than lithium alloys, due to high structural stability and lower volumetric expansion [END 00].

Anodic systems based on metal alloys (Sn, Pb, Sb, Al and Zn) [BRA 12] have also been studied. The most common negative electrode is the tin-based electrode for which $Li_{4,4}Sn$ is the final phase after lithiation. The theoretical gravimetric capacity is equal to 959.5 mAh/g [CHE 13]. However, the main issue with these electrodes is the large volume variation, and consequently. the high stresses during the lithiation process. Fractures occur rapidly and the battery life is short [CHE 13]. Composite negative electrodes based on silicon and carbon were tested. Carbon is used to decrease volume expansion of the Si-based electrode during lithiation [CHE 13].

1.1.4. *Conductive additive*

VGCF, or Vapor Grown Carbon Fiber, is a carbon in the form of microfibers (Figure 1.3), obtained by Chemical Vapor Deposition (CVD). As we can see in a Scanning Electron Microscope (SEM) image, VGCF is organized in a ball of fibers measuring several microns in length for 150 nm in diameter.

Figure 1.3. *SEM image of VGCF*

1.1.5. *Formulation of electrodes*

In an “all-solid-state” battery, the use of a solid electrolyte involves the addition of a non-negligible fraction of this electrolyte into the composition of the positive and negative electrodes. In the literature, the composite electrodes have at least 30% by weight of electrolyte, which results in a considerable "dead" volume in the battery. Since the addition of electrolyte takes place to the detriment of the active material and results in a reduction in the capacity delivered by the composite, and therefore, a reduction in the mass energy density of the battery, it is necessary to optimize the composition of the composite electrodes as well as the whole system in order to limit the proportion of solid electrolyte as much as possible. An alternative is to coat the active material with an electrolyte layer, which has in some cases made it possible to produce an electrode containing 90% mass of active material.

1.2. Shaping methods of all-solid batteries

In order to obtain good solid/solid interfaces, it is essential to adapt the shaping technique to the specific characteristics of the chosen materials. Sulfur-based electrolytes can be easily deformed and allow for the possibility of forming compact stacks by simple cold pressing, whereas oxides generally require densification at high temperatures [LAL 15].

1.2.1. *Assembly by cold pressing*

The main advantage of this technique lies in the ability of the electrolyte to promote good solid/solid interfaces and good conductivity by simple compression without resorting to high temperatures. The batteries are assembled by cold pressing in 10 mm diameter matrices, made of polycarbonate according to the steps described in Figure 1.4.

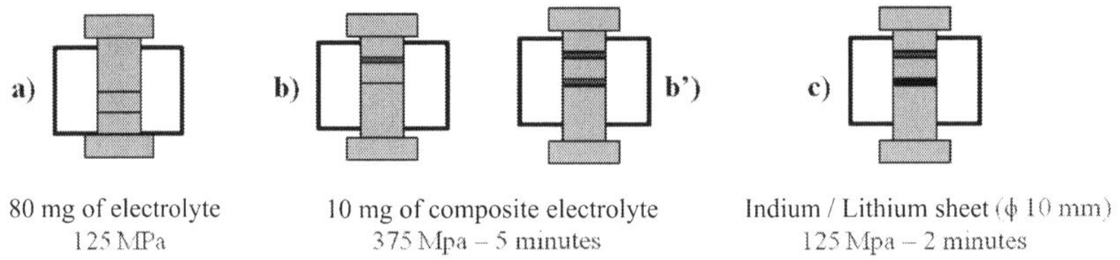

Figure 1.4. *Shaping of an "all-solid-state" battery. For a color version of this figure, see www.iste.co.uk/viallet/batteries.zip*

A half-cell is formed in three steps:

First step (Figure 1.4(a)): 80 mg of electrolyte/separator mixture is deposited in the matrix on the smaller of the two plungers and then distributed homogeneously over the entire surface of the plunger. The second plunger is then introduced into the die and a pressure of 125 MPa is applied for a few seconds. The purpose of this first step is to provide a mechanical resistance to the electrolyte.

Second step (Figure 1.4(b) and (b′)): 10 mg of composite electrode is then deposited onto the side of the smallest plunger. The plunger will then be gently turned in one direction and then in the other, in order to allow an optimal distribution of the electrode over the entire surface of the electrolyte. The second plunger is then introduced into the matrix and a pressure of 375 MPa is applied for 5 min (Figure 1.4b). In the case of a complete cell, the negative electrode is added in the same way to the other side and its assembly stops there (Figure 1.4b′).

Third step (Figure 1.4(c)): For the study of the half-cells, a sheet of lithium or indium is then introduced onto the side of the largest piston. In the case of indium, a pressure of 125 MPa is then applied for two minutes. With regard to lithium, applying pressure causes creeping of the counter electrode along the battery walls leading to its short-circuiting, it is therefore simply deposited on the electrolyte. The assembly is then placed in a device which includes screws to ensure a slight pressure during cycling (Figure 1.5(a) and (b)). Since the electrolytes are very sensitive to air and moisture, the device is placed in a sealed chamber filled with argon (Figure 1.5(c)).

Figure 1.5. *Diagram a) and photograph b) of the galvanostatic cell developed for the cycling of all-solid batteries placed in an argon glove box in a hermetic enclosure c). For a color version of this figure, see www.iste.co.uk/viallet/batteries.zip*

The problem with applying pressure during battery operation is a great concern for this type of design. Indeed, compression of the electrolyte is necessary to obtain good conductive properties and good interfaces (Figure 1.6(a)). Most studies, however, do not discuss this point which is a major obstacle for an application. However, Ito *et al.* [FLE 13] demonstrated the feasibility of a prototype $S/Li_2S–P_2S_5/Li$ of 1 Ah by conventional coating techniques familiar to the field of conventional batteries and without maintaining pressure during operation (Figure 1.6(b)).

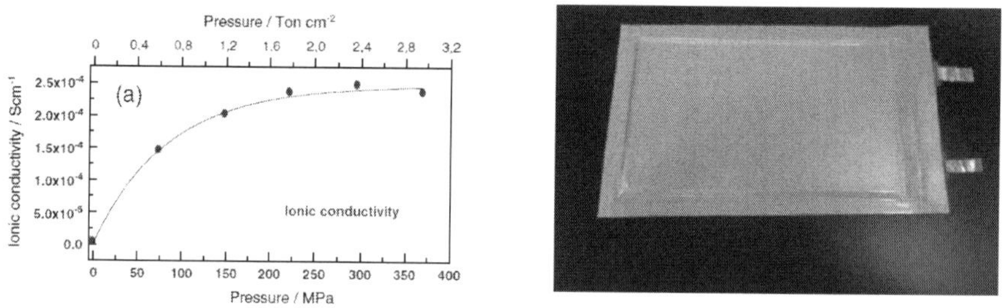

a) b)

Figure 1.6. *a) Influence of applied pressure on the ionic conductivity of Li2S–P2S5 (80:20) [FLE 12] and b) LiNi0.8Co0.15Al0.05O2/Li2S–P2S5/Li prototype of 1 Ah obtained by coating [FLE 13]. For a color version of this figure, see www.iste.co.uk/viallet/ batteries.zip*

1.2.2. *Design by high temperature sintering*

A key point for obtaining good solid/solid interfaces is the compactness of the battery. Whereas sulfides are readily compressible in the cold, oxides require very high temperatures (> 500 °C) to achieve sufficient compactness and good properties [LAL 15]. The most common approach in the literature is to densify the electrolyte by annealing at high temperature in order to obtain a pellet on which the electrodes are then deposited. The deposits are made by silkscreen printing [AGR 08, PAR 13, EVA 87], cathode sputtering [QUA 11] or coating of sol-gel followed by hot pressing [BRU 92] which makes it possible to obtain electrodes with a thickness of 5–10 µm.

This type of approach makes it possible to validate a positive electrode/ electrolyte/negative electrode system but does not tend to provide a viable means of manufacturing a monolithic "all-solid-state" battery since the electrode layers remain too thin. Current depositing techniques do not make it possible to achieve sufficient electrode thicknesses. It is therefore necessary to turn to innovative densification techniques applicable to the field of batteries.

The application of the Spark Plasma Sintering (SPS) process (or flash sintering process) to the manufacture of "all-solid-state" batteries is still very recent, the first publication being by Aboulaich *et al.* in 2011 [SMI 03]. Since then, only a few studies have been carried out, as we shall see later. Manufacturing is carried out in two stages. The first consists of stacking the layers of the battery in the powder form and in cold compaction (without the anode if it is lithium/sodium). The second is the SPS sintering of the battery (or half-battery) at high temperatures for a few minutes. If the battery uses a negative lithium metal electrode, it is added at the end of sintering.

The use of the SPS technique makes it possible to obtain very compact "all-solid" batteries with very good interfaces even in composite electrodes, which can reach thicknesses of more than 500 μm.

2

Solid Ionic Conductors

2.1. Introduction

Ionic conductive materials and their application as electrolytes in "all-solid-state" batteries have been intensively developed over the past 40 years. Li- and Na-ion conductors have been studied in a wide range of research fields ranging from ceramics to polymers to glass and glass-ceramics.

Inorganic solid electrolytes, unlike organic electrolytes, whether liquid or polymeric, have the advantage of being non-flammable and with no risk of leakage when used in a battery [TAR 01]. In addition, a mechanism for diffusion of lithium ions at the solid electrolyte/electrolyte interface that consumes less energy would theoretically lead to higher cycling rates in "all-solid-state" batteries than in conventional batteries using a liquid electrolyte [CHI 12]. Indeed, the use of a lithium salt dissolved in an organic solvent involves the desolvation of this solvent. This desolvation, necessary for the charge transfer to the electrolyte/electrode interface, requires a considerable amount of energy (~65 kJ/mol) which will limit the diffusion of the Li^+ ion and its intercalation/deintercalation during cycling (Figure 2.1) [HAY 13]. In the case of an inorganic solid ionic conductor, due to the absence of the desolvation mechanism, the energy cost of the charge transfer will be equivalent to, or even less than, that of lithium conduction (~40 kJ/mol) and will make it possible to achieve much faster regimes.

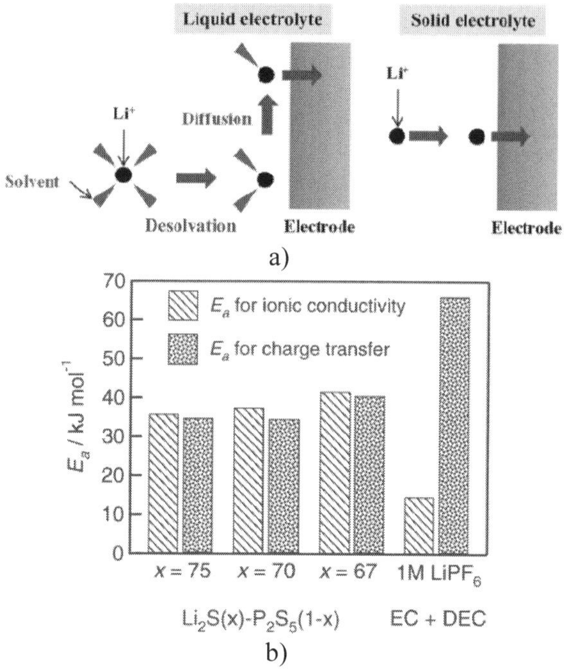

Figure 2.1. *Diagram showing the charge transfer and ion conductivity mechanisms at the electrolyte/electrode interface (a) and the associated activation energies (b). For a color version of this figure, see www.iste.co.uk/viallet/batteries.zip*

However, this phenomenon is only true in the case where the contacts between the electrode and the electrolyte are optimal, which is the case with sulfur-based electrolytes, because of their high deformability [XIE 11] in the case of shaping by cold pressing. In the case of harder ceramic electrolytes with a high sintering temperature (required for good conductivity) [GEI 11], new shaping techniques had to be developed such as assembly by Spark Plasma Sintering (SPS) or Hot Pressing (HP) or Pulsed Laser Deposition (PLD) electrode deposition to optimize interfaces and prepare efficient "all-solid-state" batteries [ABO 11, OHT 12].

In this chapter, we will first present the conduction mechanisms in crystallized and amorphous materials and we will then discuss various electrolytes, ceramics, glass and glass-ceramics, and finally, the "all-solid-state" batteries in which some have been used. The conductivity as a function of temperature of representative electrolytes is shown in Figure 2.2 [KAM 11].

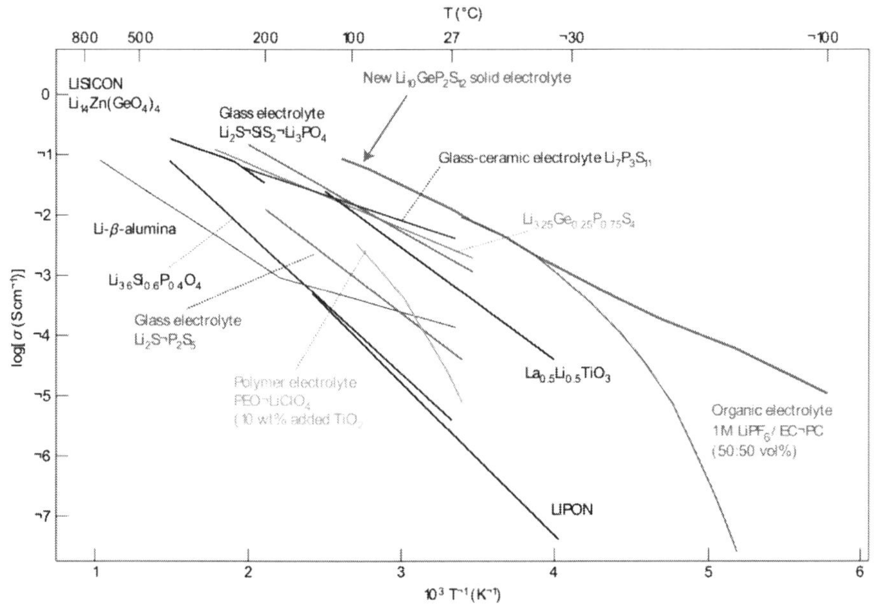

Figure 2.2. *Summary of the ionic conductivities of the main solid electrolytes in the literature. According to Kanno et al. [KAM 11]. For a color version of this figure, see www.iste.co.uk/viallet/batteries.zip*

2.2. Solid lithium-ion conductors

This section presents an overview of various ceramic, glass and glass-ceramic systems studied for their ionic conductivity properties. This overview is not intended to be exhaustive and the results presented are those considered to be the most representative of their category or those which offer a real advance in the field of solid electrolytes.

2.2.1. *The Garnets*

The structural family of the Garnets has been studied extensively since the beginning of the 20th Century [BOE 13, SMI 03], and it was in 2003 that Weppner and his team demonstrated the ionic conduction properties of this family of materials, in particular the compositions $Li_5La_3M_2O_{12}$ (M= Ta, Nb, Bi, Sb) [THA 03, MUR 07a, MUR 07b], whose general structure is shown in Figure 2.3 [THA 05b]. The garnet structure is organized according to a $[La_3La_2M_{12}]^{5-}$ skeleton consisting of LaO_8 cubes and MO_6 octahedra.

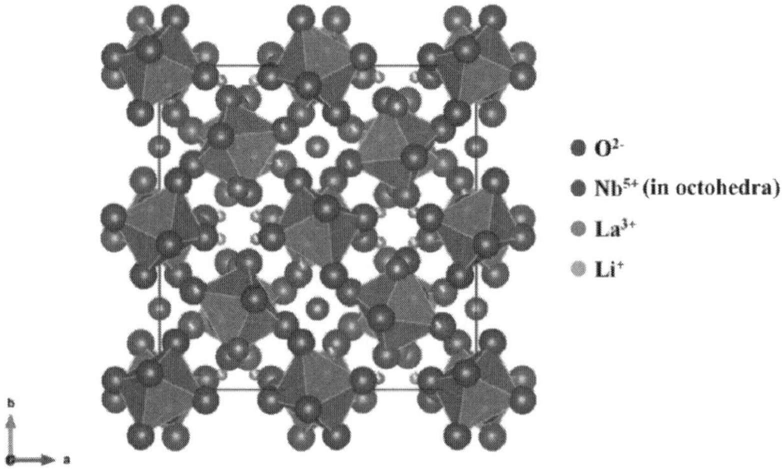

Figure 2.3. *The crystallographic structure of the compound $Li_5La_3M_2O_{12}$ (M = Nb, Ta) (a = 12.797 Å for Nb and 12.804 Å for Ta; space group $I2_13$). For a color version of this figure, see www.iste.co.uk/viallet/batteries.zip*

Like NASICON compounds, which we will see later, garnets adapt perfectly to multiple substitutions on the site of lanthanum or metal M and their properties evolve with them. Thus, the La^{3+} ion can be substituted by lower valence ions, such as Ba^{2+}, Sr^{2+} and Ca^{2+} [THA 05a], the charge being counterbalanced by an increase in the number of lithium ions. A conductivity of 4×10^{-5} S/cm at room temperature could be observed in the compound $Li_{6,25}La_{1,75}Ba_{1,25}Ta_2O_{12}$ instead of 10^{-6} S/cm for $Li_5La_3Ta_2O_{12}$ [BAT 95]. The substitution of Nb^{5+} and Ta^{5+} ions with In^{3+} ions made it possible to achieve a conductivity of 1.8×10^{-4} S/cm at 50 °C in the compound $Li_{5,5}La_3Nb_{1,75}In_{0,25}O_{12}$ [THA 06a]. The garnet $Li_7La_3Zr_2O_{12}$ has also been the subject of much research. Indeed, it has been shown that the diffusion of aluminum into the structure during a heat treatment at 1200 °C makes it possible to stabilize the high-temperature cubic phase at ambient temperature, and thus, to obtain a conductivity of 10^{-3} S/cm at 298 K coupled to an electrochemical stability of 0 at 5 V with respect to lithium [GEI 11].

Conduction in garnets is strongly dependent on the distribution of lithium ions in octahedral and tetrahedral sites during heat treatment [CUS 06, VAN 07]. Indeed, the Li^+ sites located in the octahedral sites are mobile, whereas the tetrahedral sites act as traps preventing their mobility.

2.2.2. *The NASICON $A_xMM'(XO_4)_3$ structure*

The NASICON (NA-Super-Ionic-CONductor) crystallographic structure was identified in 1968 by Kierkegard *et al.* [HAG 68] in the compounds $NaA_2(PO_4)_3$ (A=Ge, Ti and Zr). In 1976, Goodenough and Hong demonstrated ionic conduction properties that were substantially comparable to those of beta-alumina β-Al_2O_3, thus calling them Sodium Superconductors [GOO 76].

The NASICON structure can be described as a $[A_2P_3O_{12}]$ skeleton consisting of AO_6 octahedra and PO_4 tetrahedra sharing their vertices (Figure 2.4) [THA 05c]. This skeleton forms a 3D network of interconnected ionic conductive chains and has two types of interstitial positions in which the Na^+ conductive ions are distributed.

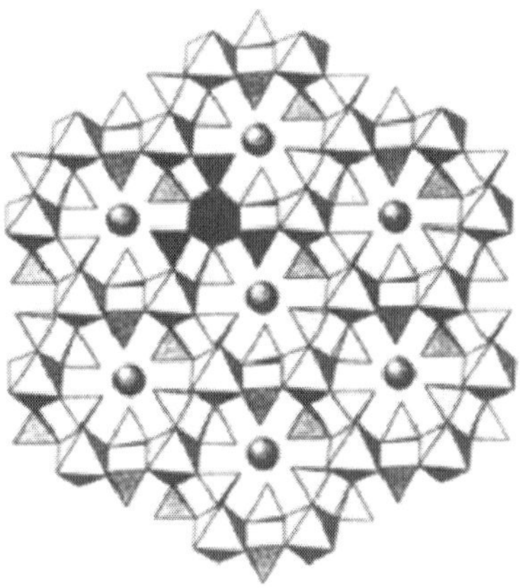

Figure 2.4. *Projection according to axis c of groups $[A_2(XO_4)_3]$ of the NASICON structure [MAT 10]*

The main characteristic of NASICON is that their structure can withstand significant local changes in composition and numerous substitutions on the A^{n+}, $M^{n'+}$ and $X^{n''+}$ sites, modifying their properties, making them compounds of choice. Among the multiple possible substitutions within the NASICON structure, that of the Li^+ ion is the most interesting. These compounds of general formula $LiA'_{2-x}{}^{IV}A''_x{}^{IV}(PO_4)_3$ see their structural and electrical properties largely influenced by the size of the ions A and A′. In the early 1990s, Aono *et al.* presented a new Li-NASICON family of the formula $Li_{1+x}Ti_{2-x}M_x(PO_4)_3$ (M = Al, Ge, In, Sc) in which the substitution of the Ti^{4+}

ion with a trivalent ion leads to a significant increase in conductivity (multiplied by 3) due to the reduction in porosity, the increase in the number of lithium in the structure and the decrease in the cell parameters [AON 89, AON 90, ARB 07]. Within this family of materials, the compound $Li_{1,3}Al_{0,3}Ti_{1,7}(PO_4)_3$ (LATP) exhibits the greatest ionic conductivity reported, 3×10^{-3} S/cm at 298 K.

The very high ionic conductivity of NASICON naturally makes them excellent candidates as electrolytes in "all-solid-state" batteries. When we refer to a battery, we expect electrochemical stability over a wide voltage range. Indeed, NASICON are stable with respect to metallic lithium and sodium provided that no reducible transition element is present in the structure. Thus, LATP, despite its excellent conductivity at room temperature, exhibits instability below 2.8 V versus Li^+/Li, which rules it out for many applications. In addition, Delmas and his team demonstrated the use of $LiTi_2(PO_4)_3$ and $NaTi_2(PO_4)_3$ electrode materials with completely reversible ion insertion/disintegration [DEL 88]. This instability at low voltages could be solved by the substitution of titanium with germanium [CRE 99] which allows the stability window to be widened up to 1.8 V versus Li^+/Li.

Recently, Goodenough and his team reported a new NASICON compound that has no elements that are likely to reduce [CED 11]; $LiZr_2(PO_4)_3$ is expected to be more stable than titanium compounds due to stability of the Zr^{4+} ion. This material, which has a triclinic structure at room temperature, exhibits a transition toward a rhombohedral structure, which is more conductive, above 50 °C. The substitution of Zr^{4+} with Ca^{2+} makes it possible to introduce more Li^+ into the structure and to stabilize the rhombohedral phase at room temperature. The compound $Li_{1,2}Zr_{1,9}Ca_{0,1}(PO_4)_3$ has a conductivity of 1.2×10^{-4} S/cm at 298 K [XIE 11, CAS 17].

2.2.3. *The compounds LISICON and Thio-LISICON*

2.2.3.1. *LISICON*

The first LISICON (LI-Super-Ionic-CONductor) to be studied was $Li_{14}ZnGe_4O_{16}$ (or $Li_{2+2x}Zn_{1-x}GeO_4$). It has a conductivity of 0.125 S/cm at 300 °C, a value close to that of $Na_3Zr_2P_1Si_2O_{12}$ [BRU 84, BRU 80, ALP 79]. This compound, like NASICON, is based on a solid solution of two less conductive compounds, Li_4GeO_4 and Zn_2GeO_4. The LISICON family is isostructural of γ-Li_3PO_4 (Figure 2.5) and consists of a 2D network of unconnected PO_4 tetrahedra [YAK 97].

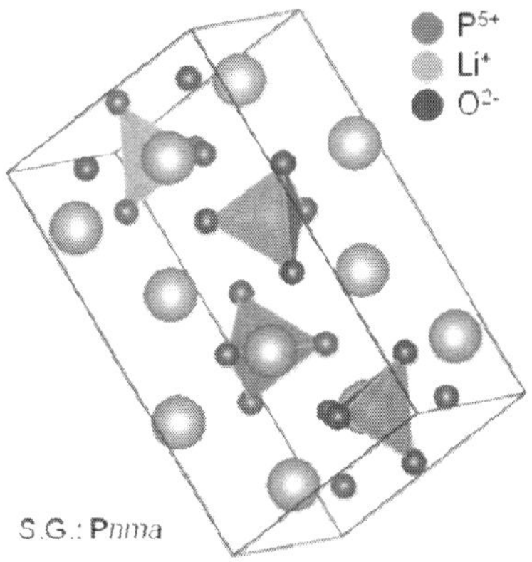

Figure 2.5. *Crystal structure of the compound γ-Li_3PO_4. (a = 10.490(3) Å, b = 6.120(2) Å and c =4.9266(7) Å; space group Pnma). For a color version of this figure, see www.iste.co.uk/viallet/batteries.zip*

Another property that LISICON shares with NASICON is their ease of accommodating substitutions. Thus, these systems have been described through numerous binary compounds such as Li_4XO_4–Li_3YO_4 or Li_4XO_4–Li_5AlO_4 (X = Si, Ge, Ti and Y = P, As, V, Cr) [MAS 95a, MAS 95b, ROB 97]. Despite their relatively low ionic conductivity (10^{-5}–10^{-6} S/cm at room temperature), these exhibit a high thermodynamic stability versus lithium, even in the presence of molten lithium, and excellent conductivity at high temperature (300 °C), which led to considerable research in the 1980s to 1990s for application as electrolytes in thermal cells ($Li_{3,6}Ge_{0,6}V_{0,4}O_4$ has been used by the US military). Others, such as $Li_{3,4}Si_{0,4}V_{0,6}O_4$ and $Li_{3,6}Si_{0,6}P_{0,4}O_4$, have been used as electrolytes in micro-batteries [ROB 97].

2.2.3.2. *Thio-LISICON*

The Thio-LISICON family, sulfur derivatives of LISICON, was introduced by Kanno and his team [KAN 01] in the early 2000s, inspired by Ahn's works on Li_4SiS_4 [MUR 02a, MUR 02b, AHN 89]. It is based on the substitution of the O^{2-} ion by a larger and more polarizable ion, the S^{2-} ion, in order to increase the mobility of Li^+ ions. The use of sulfur raises the need to be handled under a controlled atmosphere, because of its sensitivity to air and moisture. Thio-LISICON are also isostructural of the compound γ-Li_3PO_4 and have the same predisposition to substitutions. As shown in Figure 2.6, the $O^{2-} \Leftrightarrow S^{2-}$ substitution makes it possible, at best, to increase the ionic conductivity by two orders of magnitude (i.e. Li_3PCh_4–Li_4SiCh_4 system, Ch = O, S) and even obtain compounds competing strongly with conventional organic liquid electrolytes (i.e. $LiPF_6$ + Carbonates: ~1×10^{-2} S/cm).

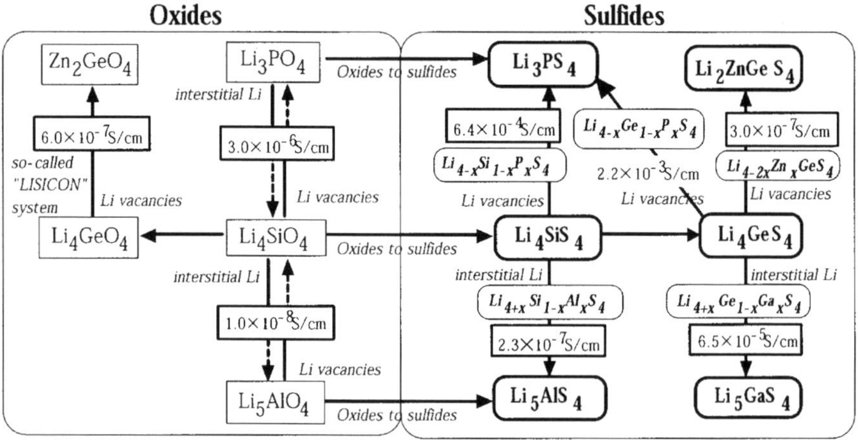

Figure 2.6. *Comparative table of the conductivity of LISICON and their sulfur counterparts [KAN 01]*

The solid solution Li_4GeS_4–Li_3PS_4 has been studied most extensively because it has the greatest interest from a structural and electrochemical point of view [KAN 01]. As shown in Figure 2.7(a), the study of the lattice parameters of the $Li_{4+x}Ge_{1-x}P_xS_4$ compound as a function of the substitution rate has made it possible to demonstrate three distinct domains corresponding to three trends in the evolution of these parameters. These three regions, I ($0 \le x \le 0.6$), II ($0.6 < x < 0.8$) and III ($0.8 \le x \le 1$), were studied by Hayashi *et al.* under the term Thio-LISICON analog I, II or III [HAY 03b]. Measuring the ionic conductivity as a function of the substitution rate x shows that this increases to a maximum of 2.2×10^{-3} S/cm for the composition $Li_{3,25}Ge_{0,25}P_{0,75}S_4$ ($x = 0.75$, region II), before decreasing for x rates in region III (Figure 2.7(b)). The compound $Li_{3,25}Ge_{0,25}P_{0,75}S_4$ exhibits excellent electrochemical stability with respect to Li (Figure 2.7(c)) and will subsequently give rise to several applications as an electrolyte in all-solid-state Li-ion batteries

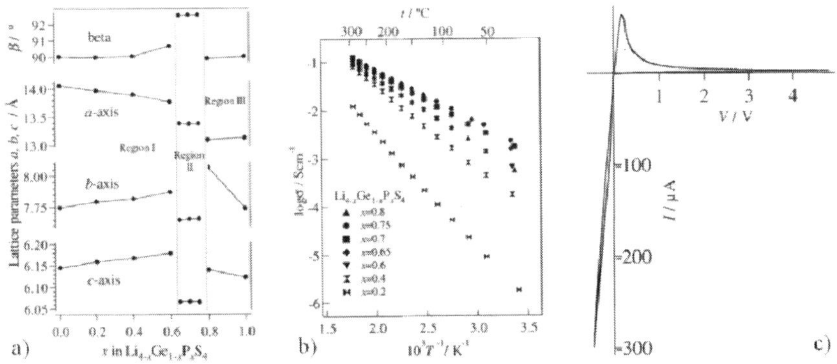

Figure 2.7. *a) Cell parameters and b) ionic conductivity at temperatures between 25 and 300 °C for $Li_{4+x}Ge_{1-x}P_xS_4$ as a function of x. c) Electrochemical stability of $Li_{3.25}Ge_{0.25}P_{0.75}S_4$ recorded at 1 mV/s in a Li/ $Li_{3.25}Ge_{0,25}P_{0.75}S_4$/steel cell [KAN 01]*

2.2.3.3. *$Li_{10}GeP_2S_{12}$, quadratic LGPS structure*

Recently, Kanno and his team revolutionized the field of ionic conductors by presenting a solid electrolyte whose ionic conductivity is higher than that of a conventional liquid electrolyte [KAM 11]. The compound $Li_{10}GeP_2S_{12}$ (or $Li_{4-x}Ge_{1-x}P_xS_4$ with $x = 0.66$) has a conductivity of 12×10^{-3} S/cm at 298 K with an activation energy of 24 kJ/mol as well as stability up to 5 V. These conduction properties are equivalent to those exhibited by an organic liquid electrolyte and in addition, as shown in Figure 2.8(a), this material exhibits very interesting conductivity properties under extreme conditions (0.1 S/cm at +100 °C and 10^{-5} S/cm at −100 °C) where it largely supersedes liquid electrolytes (Figure 2.2).

$Li_{10}GeP_2S_{12}$ has a three-dimensional structure (space group $P4_2/nmc$, 137) based on tetrahedra ($Ge_{0.5}P_{0.5}S_4$), PS_4, LiS_4 and LiS_6 octahedra (Figure 2.8(b)), with the diffusion of lithium ions occurring along the c axis but also in the planes ab [MO 11, MAS 11].

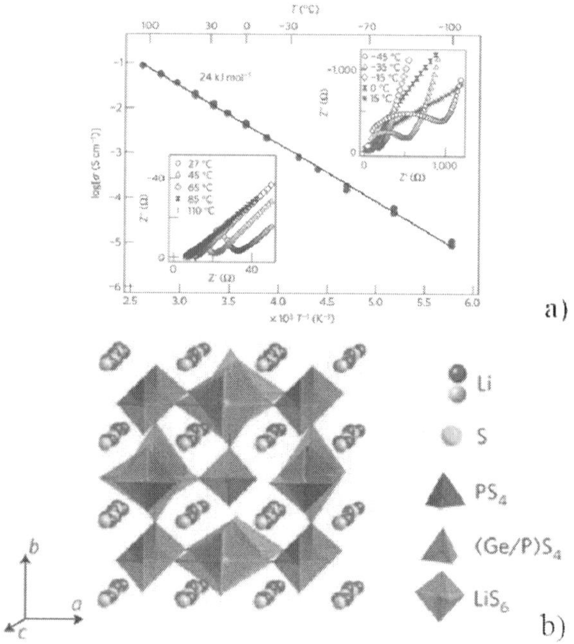

Figure 2.8. *a) Ionic conductivity of $Li_{10}GeP_2S_{12}$ as a function of temperature between –45 and +110 °C. b) Crystallographic structure of $Li_{10}GeP_2S_{12}$ [KAM 11]. For a color version of this figure, see www.iste.co.uk/viallet/batteries.zip*

This material immediately aroused the interest of many groups, notably that of Ceder at MIT (Massachusetts Institute of Technology) [MO 11]. DFT (Density Functional Theory) and molecular dynamics calculations showed two oxidation (decomposition into GeS_2, S and P_2S_5) and reduction (decomposition into Li_2S, Li_3P and $Li_{15}GeS_4$) phenomena at the surface of the compound at high and low voltage, respectively. It is this decomposition which allows the compound $Li_{10}GeP_2S_{12}$, due to the formation of a passivation layer, to remain stable up to 5 V.

However, the use of $Li_{10}GeP_2S_{12}$ as a solid electrolyte still poses two problems. First of all, germanium is extremely expensive (1855 €/kg on June 28, 2013, source: http://bourse.lesechos.fr), which rules it out for industrial applications. Nevertheless, in 2012, Toyota introduced an electric scooter equipped with an "All-Solid-State Battery" prototype made up of 18 cells in parallel with a voltage of 28 V, as we shall present later [SHI 12]. The second problem of this compound is the use of sulfur which, when exposed to air, can release H_2S, hydrogen sulfide, toxic at high concentrations.

In order to overcome these two issues, Ong *et al.* carried out a DFT study on the family of compounds $Li_{10\pm x}MP_2Ch_{12}$ (M = Ge, Si, Sn, P or Al and *Ch* = O, S and Se) and determined their ionic conductivities and activation energies, which are shown in Table 2.1 [ONG 13].

Compound	E_a (eV)	Conductivity (mS/cm)
$Li_{10}GeP_2S_{12}$	0.21 ± 0.04	13
$Li_{10}GeP_2O_{12}$	0.36 ± 0.05	0.03
$Li_{10}GeP_2Se_{12}$	0.19 ± 0.04	24
$Li_{10}SiP_2S_{12}$	0.20 ± 0.03	23
$Li_{10}SnP_2S_{12}$	0.24 ± 0.03	6
$Li_{11}AlP_2S_{12}$	0.18 ± 0.06	33
$Li_9P_3S_{12}$	0.26 ± 0.09	4

Table 2.1. *Overview of ionic conductivities of $Li_{10\pm x}MP_2Ch_{12}$ compounds [ONG 13]*

The substitution of germanium with silicon or aluminum leads to a conductivity that is considerably higher than that of $Li_{10}GeP_2S_{12}$ (i.e. multiplied by 2 for $Li_{10}SiP_2S_{12}$) at a much lower material cost. The incredible conductivity of 33 mS/cm for $Li_{11}AlP_2S_{12}$ (three times greater than that of a liquid electrolyte) can be attributed to the increase in the concentration of lithium ions in the structure. With regard to the substitution of the anion, the use of an even larger and more polarizable ion such as selenium leads to an increase in conductivity, but its toxicity rules it out for industrial applications.

2.2.4. *Ion conductive glass and glass-ceramics*

2.2.4.1. *The first vitreous electrolytes*

Oxide-based conductive glasses, such as $Li_2O–SiO_2–Al_2O_3$ [ROG 00], have been studied for many years, but their low conductivity at ambient temperature has drastically reduced their interest and they have given way to glass and glass-ceramics based on sulfur.

In the 1980s, several French groups held the top position with regard to sulfur-based ionic conductive glass and were to be the source of many systems that would be developed and improved by teams from around the world. In Besançon, Mercier *et al.* initiated work on the binary compound

$Li_2S–P_2S_5$ [KAN 83], research that was recommenced almost 20 years later by Hayashi *et al.* in Osaka and which led to the synthesis of an electrolyte allowing the manufacture of high-performing all-solid-state batteries. The compound $60Li_2S–40P_2S_5$ (by percentage mass) obtained by quenching the glass between two rollers has a conductivity of 10^{-4} S/cm at 298 K. This conductivity can be increased to a maximum of 10^{-3} S/cm by adding to lithium iodide LiI to the system, which is a well-known network former, up 45 wt%, as shown in Figure 2.9.

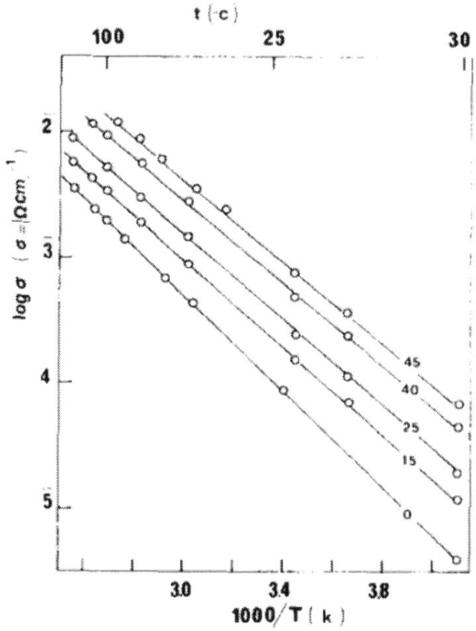

Figure 2.9. *Influence of temperature on ionic conductivity of the system (1–x) ($60Li_2S–40P_2S_5$) –xLiI [KAN 83]*

However, this increase in conductivity is to the detriment of its electrochemical stability, which does not exceed 2.6 V versus Li^+/Li [KAN 83]. In 2013, the $Li_2S–P_2S_5–LiI$ system was found to reach a maximum of 2×10^{-3} S/cm (at 298 K) in the glass-ceramic $95(80Li_2S–20P_2S_5)$ – 5 LiI [UJI 13].

Ménétrier *et al.* in Bordeaux were interested in the $Li_2S–B_2S_3–LiI$ system which has a conductivity of 10^{-3} S/cm at 298 K [WAD 83, MEN 84]. In Montpellier, Pradel and Ribes were working on the system Li_2S–(Ge, Si)S_2–

LiI [RIB 80, PRA 86] with a slightly lower conductivity of 8×10^{-4} S/cm. At the international level, the bibliography of these glass compounds was predominantly dominated by Kennedy *et al.* of the University of Santa Barbara until the early 1990s. Inspired by the French teams, they presented very complete conductivity studies on the electrochemical and structural properties of many systems such as Li_2S–P_2S_5–SiS_2 [KEN 88], Li_2S–SiS_2–LiX (X= Br, Cl or I) [KEN 87] or Li_2S–P_2S_5–B_2S_3 [ZHA 90].

2.2.4.2. *A new synthesis route*

In the early 1990s, teams from Japanese laboratories in Osaka, including Professors Tatsumisago and Minami, were interested in the Li_2S–P_2S_5 and Li_2S–SiS_2 systems. The first vitreous sulfur-based electrolyte studied in Japan was an oxysulfide in the Li_2S–SiS_2 system by Kondo and Takada [AOT 94, KON 92]. Li_3PO_4–Li_2S–SiS_2 and Li_2SO_4–Li_2S–SiS_2 have a lower conductivity than their LiI counterparts, 6.9×10^{-4} S/cm at 298 K, but are stable up to 5 V versus Li^+/Li. This conductivity was then optimized up to 1.4×10^{-3} S/cm, allowing for the assembly of the first "all-solid-state" battery using high-potential positive electrode materials such as $LiCoO_2$ [TAK 96].

After collaboration with Kondo on Li_3PO_4 glass, Tatsumisago *et al.* added LISICON Li_4SiO_4 and Li_4GeO_4 to the Li_2S–SiS_2 system [HIR 95, TAT 97, HAY 99]. The Li_4SiO_4–Li_2S–SiS_2 system is the most promising with a conductivity of 10^{-3} S/cm at 298 K and excellent chemical and electrochemical stability with respect to lithium. This electrolyte will also be tested and validated in an "all-solid-state" battery as shown in Chapter 3.

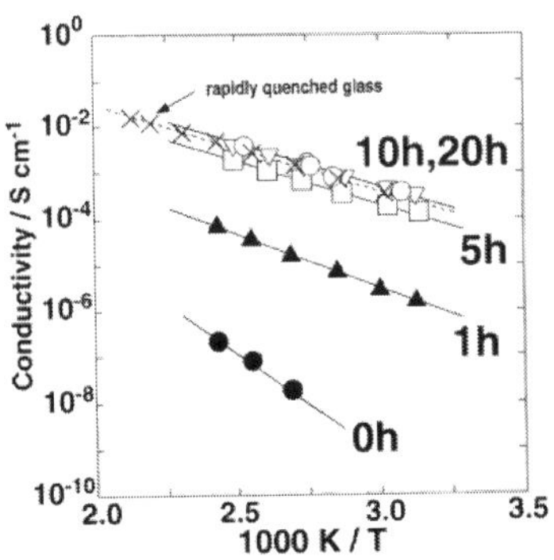

Figure 2.10. *Impact of the grinding time on the ionic conductivity of Li_2S–SiS_2–Li_4SiO_4*

In the early 2000s, Morimoto *et al.* chose to replace the traditional quenching synthesis of molten glass with a new, more energetic method: mechanosynthesis [MOR 99, MOR 00]. This technique, based on the energy grinding of Li_2S, SiS_2 and Li_4SiO_4 precursors in an alumina jar with similar beads, makes it possible to obtain, in only 10 hr (triangles in Figure 2.10), glasses with exactly the same properties as glasses obtained by quenching (cross in Figure 2.10) [TAT 00]. This new, simpler synthesis route marks a turning point in the development and preparation of solid electrolytes of this family.

2.2.4.3. *From glass to glass-ceramics*

Using this new synthesis path, Professor Tatsumisago's team then began to take an interest in the Li_2S–P_2S_5 system developed by Mercier and thus initiated numerous studies on this binary compound. The study of the Li_2S–P_2S_5 glass obtained by mechanosynthesis made it possible to determine a maximum conductivity of 10^{-4} S/cm for $75Li_2S$–$25P_2S_5$ (in percentage mass), as shown in Figure 2.11(a) [HAY 01, TAT 02]. The thermal analysis curve of the amorphous powder $80Li_2S$–$20P_2S_5$ in Figure 2.11(b) shows an exothermic peak at 210 °C, which can be attributed to the phenomenon of crystallization, the glass transition being observed at 190 °C.

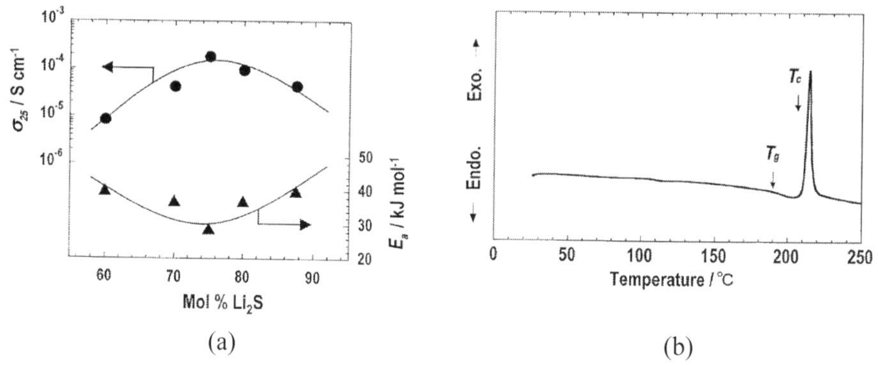

Figure 2.11. *Main results of the study of glass (100–x)Li_2S–xP_2S_5. a) Conductivity as a function of x. b) Differential thermal analysis curve of powder $80Li_2S$–$20P_2S_5$ obtained after 40 hr of mechanical grinding*

Thermal treatment of this compound beyond its crystallization temperature results in the precipitation of an iso-structural phase of $Li_{3.25}Ge_{0.25}P_{0.75}S_4$, called "Thio-LISICON analog II" [HAY 03a] which greatly contributes to the increase in conductivity, as shown in Figure 2.12(b) and (c). This new glass-ceramic has a high conductivity equal to 7.2 × 10^{-4} S/cm (at 298 K), brought to 10^{-3} S/cm by the modification of experimental conditions. In the case of $75Li_2S–25P_2S_5$, the crystallization of "Thio-LISICON analog III" does not contribute to an increase in conductivity.

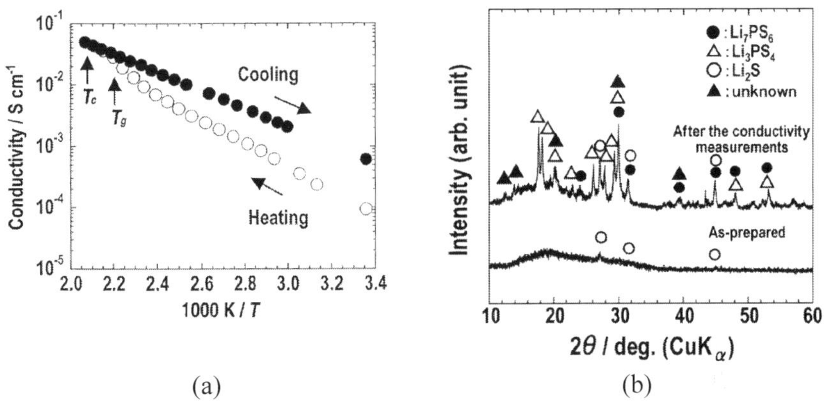

Figure 2.12. *a) Conductivity as a function of temperature for $80Li_2S–20P_2S_5$. b) X-ray diffraction pattern before and after the temperature study [TAT 02]*

Due to the similarity between the $Li_2S–P_2S_5$ based glass and Thio-LISICON, the glass-ceramic compound $Li_3PS_4–Li_4SiS_4$ was prepared by mechanical grinding followed by heat treatment, resulting in a conductivity twice that of its ceramic counterpart [JOH 58].

The difference in conductivity between a ceramic and the corresponding glass-ceramic lies in their radically opposite shapes. A glass-ceramic can be described as crystallized particles immersed in an amorphous matrix of glass of the same composition, thus ensuring a continuous network throughout the material. This vitreous matrix, easily deformable, allows optimal percolation of the more conductive crystallized particles, drastically reducing the resistance to grain boundaries. As shown in Figure 2.13, sulfur-containing glass such as in $Li_2S–P_2S_5$ (a) exhibit excellent compactness compared to ceramics such as $Li_7La_3Zr_2O_{12}$ (b), even after simple cold pressing. Moreover, this matrix can lead to a favorable interface between the electrolyte and the electrode during its use in batteries, but it can also withstand volume expansions of various electrode materials.

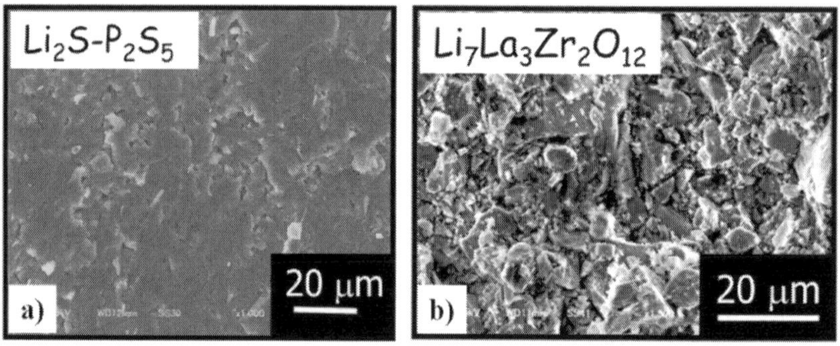

Figure 2.13. *SEM images of a transverse section of a cold-pressed Li_2S–P_2S_5 glass pellet (a) and the garnet $Li_7La_3Zr_2O_{12}$ (b) [MUR 07]*

The discovery of a "Super Cristal" $Li_7P_3S_{11}$ in the Li_2S–P_2S_5 system, Figure 2.14(a), will once again revive and widen the field of study on this binary compound [MIZ 06, YAM 07].

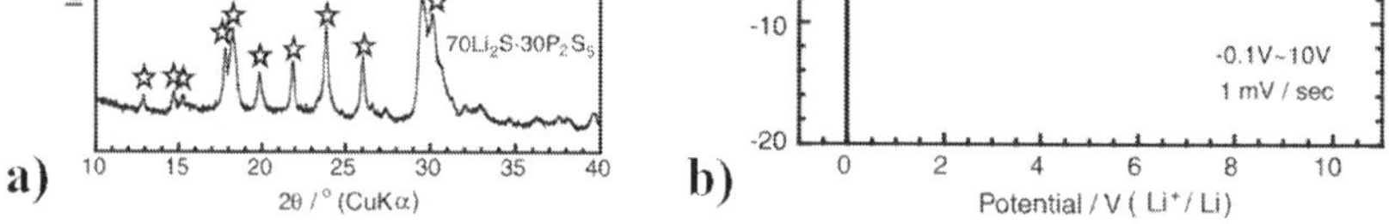

Figure 2.14. *a) Diagram of X-ray diffraction (λCu Kα) of glass-ceramics in the system (100–x) Li_2S–xP_2S_5. b) Electrochemical stability of the glass-ceramic $70Li_2S$–$30P_2S_5$ (right) [133]*

Indeed, the glass-ceramic $70Li_2S$–$30P_2S_5$ has a conductivity of 3.2×10^{-3} S/cm after precipitation of the $Li_7P_3S_{11}$ phase. Despite this very high conductivity, the glass-ceramic exhibits, as seen in Figure 2.14(b), a reactivity at approximately 2 V versus Li^+/Li due to the oxidation of the S^{2-} ions in the structure. This oxidation phenomenon could be limited by the addition of P_2S_3 [MIN 09], P_2O_5 [MIN 08, HAY 09b] or GeS_2 [MIN 10], also resulting in an increase in the ionic conductivity of the compound.

2.2.5. *The Argyrodites*

The origin of the Argyrodites dates back to the end of the 19th Century with the discovery of the compound Ag_8GeS_6 in the Himmelsfurst mines in Germany [WEI 86]. The origin of the term Argyrodite comes from the Greek Arghyrodes meaning "that contains silver". Research into these compounds only began at the end of the 1970s. Argyrodites, like garnets and NASICON, have a wide range of compositions and in 1979 Kuhs *et al.* established a general formula for this family of materials [KUH 79]:

$$A^{m+}_{12-n-x}\,B^{n+}\,Ch^{2-}_{6-x}\,X^{-}_{x}$$

with A = Cu, Ag, Li

B = Ge, Si, Al, P

Ch = O, S, Se, Te

X = Cl, Br, I

The first lithium ion-conducting Argyrodites $Li_{7-x}PCh_{6-x}X_x$ (*Ch* : O, S, Se ; X= Cl, Br and I) were presented in 2008 by Deiseroth *et al.* [DEI 08]. These compounds were obtained by a solid-state reaction of the mixture Li_2S, P_2S_5 and LiX in a tube sealed in a vacuum at 550 °C for 7 days. At the end of this heat treatment, the chlorinated and brominated compounds showed an ionic conductivity approximately 10^{-4} S/cm at ambient temperature suggesting the possible use of this new family of compounds as solid electrolytes in all-solid-state Li-ion batteries.

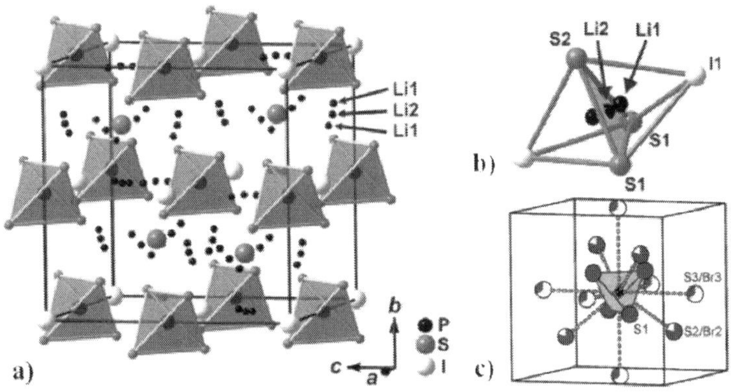

Figure 2.15. *a) Crystalline structure of Li_6PS_5I. b) Distribution of Li^+ ions in I_2S_3 polyhedra. c) S/X Substitutions (X = Cl or Br) at sites X1 and S2*

The compounds Li_6PS_5X (X= Cl, Br or I) are organized according to a Face-Centered Cubic network of anions X^- in which the tetrahedra $(PS_4)^{3-}$ occupy 100% of the octahedral sites and the S^{2-} ions occupy 50% of the tetrahedra sites (Figure 2.15(a)). The halide anions are denoted X1, the ions S^{2-} forming the tetrahedra are denoted S1 and the S^{2-} ions in the tetrahedral sites are denoted S2. The Li^+ cations are distributed in two different positions, denoted Li1 and Li2, within polyhedra formed by two anions X^-, two sulfur S1 and one sulfur S2, as shown in Figure 2.15(b). The random distribution of lithium ions is at the origin of the high ionic conductivity of these compounds, occurring by jumping from one polyhedron to another. Moreover, within a polyhedron, the simultaneous occupation of both positions is omitted. In the case of Li_6PS_5Cl and Li_6PS_5Br, two substitutions are possible between the Br^- or Cl^- ions and the S^{2-} ions at sites S2, giving rise to the notation X2, and at sites X1, giving rise to the notations S3 and X3 (Figure 2.15(c)). The substitution of S^{2-} ions with more electronegative halide ions results in a strong delocalization of the charge within the polyhedra allowing a greater mobility of the Li^+ ions and contributes to making these compounds good ionic conductors. The absence of S/I substitutions in the compound Li_6PS_5I can be explained by a significant difference in ionic rays between I^- (220 pm) and S^{2-} (184 pm) compared to anions Cl– (181 pm) and Br– (196 pm).

Since 2011, Rao *et al.* have also been interested in this new family of ionic conductors. The synthesis of phases is carried out in two stages. First, Li2S–P2S5–LiX (X = Cl, Br, I) mixtures are mechanically ground for 20 hr in an agate jar with 15 balls of the same nature 10 mm in diameter. Second, the powders obtained are heat-treated at 550 °C for 5 hr [RAO 11, RAO 12].

The X-ray diffraction diagrams show that the powders obtained after grinding are partially amorphous and only the Bragg peaks of Li_2S, P_2S_5 and LiX (X = Cl, Br, I) are observed. The peaks of Li_6PS_5Cl, Li_6PS_5Br and Li_6PS_5I only appear after heat treatment and the phases are perfectly crystallized [RAO 11]. These initial studies show that grinding steps considerably reduce the heat treatment duration (5 hr instead of 7 days for Deiseroth *et al.*).

Partially amorphous samples were obtained directly after grinding, with ionic conductivities, respectively, equal to 3.3×10^{-5}, 3.2×10^{-5} and 2.2×10^{-4} S/cm at ambient temperature with activation energies, respectively, equal to 0.38, 0.32 and 0.26 eV [139] (Figure 2.16). The ionic conductivities of Li_6PS_5Cl, Li_6PS_5Br and Li_6PS_5I, completely crystallized after heat treatment, are equal to 7.4×10^{-4}, 7.2×10^{-4} and 4.6×10^{-7} S/cm with activation energies equal to 0.11, 0.17 and 0.25 eV.

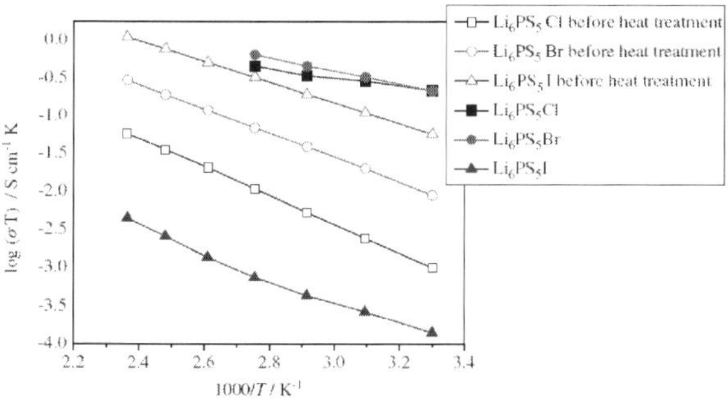

Figure 2.16. *Representation, in Arrhenius coordinates, of the ionic conductivity of Li_6PS_5Cl, Li_6PS_5Br and Li_6PS_5I before (open symbols) and after (closed symbols) heat treatment [RAO 12]. For a color version of this figure, see www.iste.co.uk/viallet/batteries.zip*

The combined analyses of high-resolution neutron diffraction and X-ray diffraction in the laboratory showed that lithium ions are randomly distributed at their sites in the three phases, and that for Li_6PS_5Cl and Li_6PS_5Br, random S^{2-}/Cl^- and S^{2-}/Br^- substitutions take place. These substitutions are not observed in Li_6PS_5I, since the I^- ions are too large. The more orderly structure Li_6PS_5I has a more moderate conductivity than the argyrodite phases with chlorine and bromine. These results confirm those obtained by Deiseroth *et al.*, regarding the positive influence of the structural disorder on ionic conductivity.

Rao *et al.* continued their research in order to better understand the influence of the preparation conditions on the conductivity of argyrodite phases [RAO 13]. The Bragg peaks of the argyrodite phases $Li_{7-x}PS_{6-x}Cl_x$ (Space group F-43 m) appear from 80 °C and the phases are predominantly crystallized at approximately 150 °C, at which temperature the Li_2S phase disappears. The peaks of LiCl only disappear at approximately190 °C. The argyrodite phase is crystallized up to 490 °C, but becomes amorphous beyond this temperature. However, if the temperature is lowered again, the phase crystallizes again.

The total Cl content denoted x was determined for different temperatures during heating to 430 °C and cooling from this value to ambient temperature (Figure 2.17(a)). Up to 190 °C, the argyrodite phase does not contain a significant amount of Cl. Beyond this temperature, the chlorine content increases until it reaches the value x = 2/3 at T = 310 °C and increases up to x = 1 at 430 °C. When cooled from 430 °C, x decreases to a value of 0.5 at 310 °C and increases slightly to stabilize at a value of 0.65 from 310 °C to ambient temperature.

The cooling curves carried out after a first heat treatment up to 150, 200, and 250 °C (Figure 2.17(b)) show that the conductivity depends on the chlorine content in phase $Li_{7-x}PS_{6-x}Cl_x$. The compound heated at 250 °C (squares on the curve), and therefore containing the most chlorine, has the highest ionic conductivity equal to 1×10^{-3} S/cm.

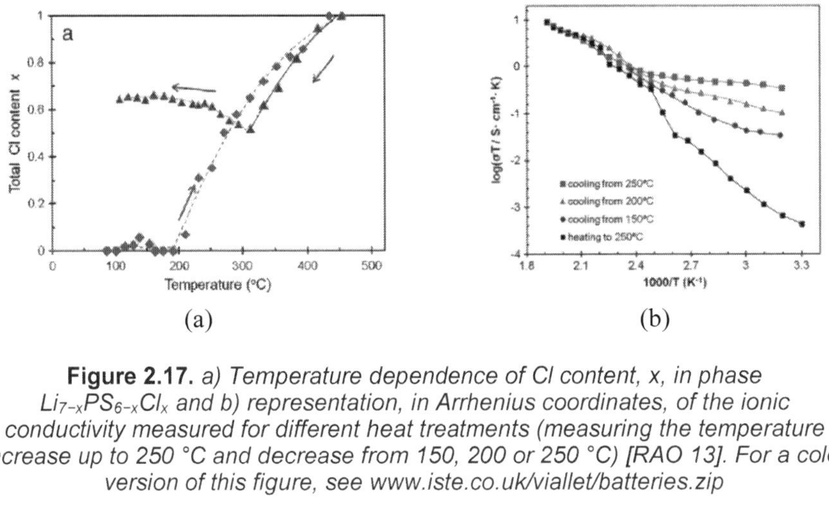

(a) (b)

Figure 2.17. *a) Temperature dependence of Cl content, x, in phase $Li_{7-x}PS_{6-x}Cl_x$ and b) representation, in Arrhenius coordinates, of the ionic conductivity measured for different heat treatments (measuring the temperature increase up to 250 °C and decrease from 150, 200 or 250 °C) [RAO 13]. For a color version of this figure, see www.iste.co.uk/viallet/batteries.zip*

At the same time, Boulineau *et al.* also developed a synthesis of Li-argyrodite phases Li_6PS_5X (X = Cl, Br, I) involving a mechanical grinding step and studied the impact of this step on the transport properties [BOU 12]. A high grinding energy ball mill (Pulverisette 7) was used with a relatively high speed of 600 revolutions per minute continuously for 20 hr. The reagents Li_2S, P_2S_5 and LiX (X = Cl, Br, I) are placed in zirconium oxide grinding jars and balls of the same nature. The samples obtained after mechanical grinding are heat treated at 550 °C for 5 hr.

The X-ray diffraction diagrams of the three Argyrodites prepared by planetary milling exhibit characteristic peaks of the compounds Li_6PS_5Cl, Li_6PS_5Br and Li_6PS_5I, that are extremely broad (Figure 2.18(a)), and a high continuous spectrum characteristic of the presence of an amorphous phase in all three cases. Studying the width of the diffraction peaks [PAT 39] determined the size of the crystallites to be at about 15 nm for the three compounds. This initial characterization shows that the Argyrodites Li_6PS_5X (X = Cl, Br or I) form spontaneously during milling, whatever the nature of the halogen, and without any heat treatment. This is called mechanosynthesis and no longer merely "mechanical milling".

The influence of the heat treatment was monitored by X-ray diffraction.

A first heat treatment carried out with a temperature ramp of 10 °C/min results in an increase in the intensity of the peaks, thus a better crystallization of the phases and a narrowing of the width of the peaks and therefore an increase in the size of the crystallites, close to ~90 nm (Figure 2.18(b)). The crystallization of Li_2S and LiCl precursors is also observed. P_2S_5, whose melting point is 288 °C, is extruded out of the pellet during processing leading to an inhomogeneous composition and the isolated crystallization of the Li_2S (●) and LiCl (♦) precursors, instead of participating in the formation of the Li_6PS_5Br phase. A faster temperature ramp of 25 °C/min causes a slight decrease in the quantity of Li_2S and a melting limit of P_2S_5.

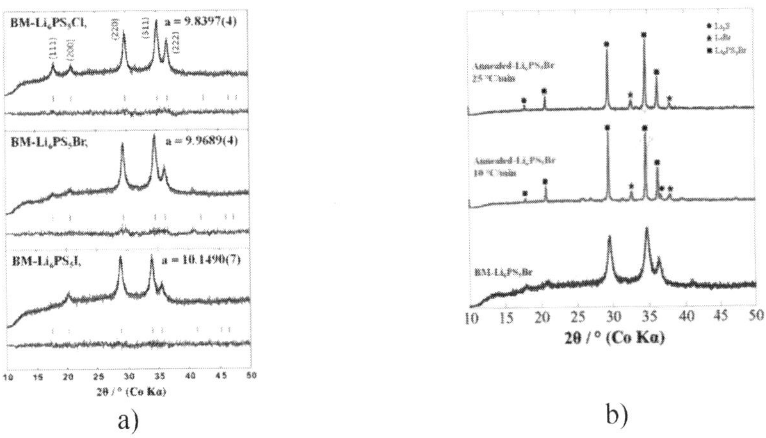

Figure 2.18. *a) XRD data of BM-Li_6PS_5X (X=Cl, Br, I) phases. b) XRD diffraction diagrams of Li_6PS_5Br phases with mechanical grinding and after heat treatment at 550 °C for 5 hr for two heating rates (10 and 20 °C/min) [BOU 12]. For a color version of this figure, see www.iste.co.uk/viallet/batteries.zip*

Since Argyrodite phases are directly formed by mechanical grinding, the impact of the duration of this grinding on the formation of the crystallized phase has been studied (Figure 2.19). For low grinding durations of 1 and 2 hr, only the Bragg peaks of Li_2S and LiCl are observed on the X-ray diffractograms (Figure 2.19(a)). After 5 hr of grinding, the Bragg peaks of Li_6PS_5Cl appear but the peaks are wide and the high continuous spectrum indicates a low crystallization. Li_2S can still be observed on the diffractogram. Between 10 and 20 hr of grinding, the crystallized Li_2S phase has virtually disappeared and the pure Li_6PS_5Cl phase is obtained.

The conductivities at ambient temperature, determined by impedance spectroscopy measurements, increased by 2×10^{-4} S/cm for 1 hr of grinding at 1.33×10^{-3} S/cm for 10 hr of grinding and decreases with longer grinding durations (Figure 2.19(b)).

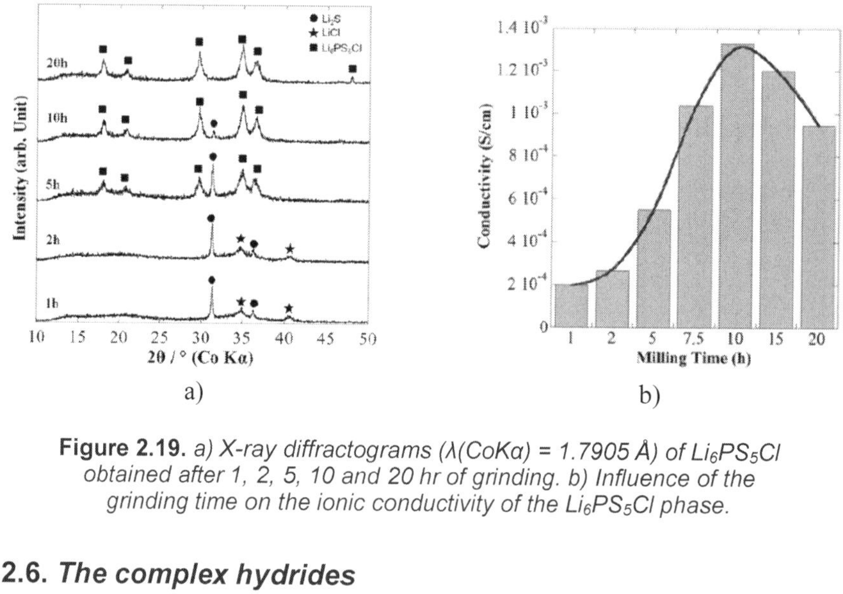

a) b)

Figure 2.19. *a) X-ray diffractograms (λ(CoKα) = 1.7905 Å) of Li_6PS_5Cl obtained after 1, 2, 5, 10 and 20 hr of grinding. b) Influence of the grinding time on the ionic conductivity of the Li_6PS_5Cl phase.*

2.2.6. *The complex hydrides*

The first and most well-known example of hydride-based ionic conductors is lithium borohydride $LiBH_4$ [MAE 09] which, at 113 °C, has an orthorhombic phase transition (low-temperature (LT) phase) toward a hexagonal phase (high-temperature (HT) phase), allowing its conductivity to increase by three orders of magnitude, that of the HT phase being 10^{-3} S/cm at 118 °C. Stabilization of the HT phase at room temperature has been extensively researched and the addition of LiI seems to have solved this problem [MIY 11]. Indeed, the compound $3LiBH_4.LiI$ (25 mol% of LiI) undergoes no phase transition between 25 and 120 °C, and has a conductivity of 10^{-4} S/cm at 298 K (10^{-8} S/cm for pure $LiBH_4$). Figure 2.20 shows the impact of the addition of lithium halide on the electrical DC conductivity of the system $LiBH_4$ [MAE 09]. Recently, Ley *et al.* also showed a conductivity of 3×10^{-4} S/cm at room temperature for the family of compounds $LiM(BH_4)_3Cl$ (M = La, Gd) [LEY 12].

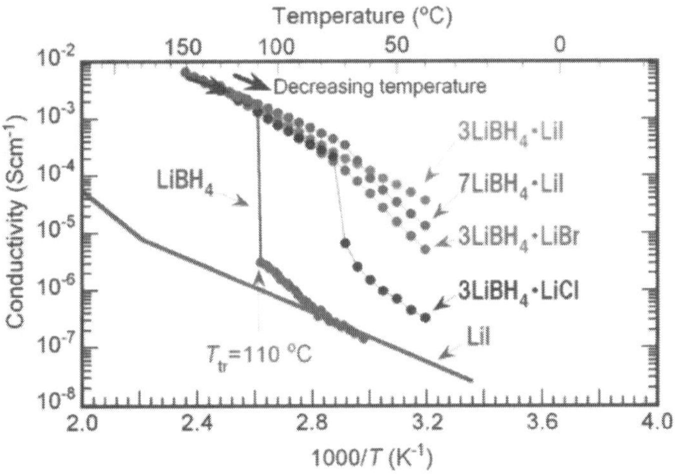

Figure 2.20. *DC electrical conductivity as a function of temperature of compounds $LiBH_4$ and LiI and $LiBH_4$–LiX composites (X = Cl, Br and I) [MAE 09]. For a color version of this figure, see www.iste.co.uk/viallet/batteries.zip*

The second family of hydrides based on $Li(NH_2)$ has also seen its conductivity increase by the insertion of LiI leading to a conductivity of 10^{-5} S/cm at 298 K for the compound $Li_3(NH_2)_2I$, instead of 3×10^{-9} S/cm [MAT 10].

Complex hydrides based on anions $(BH_4)^-$ and $(NH_2)^-$ have ionic conductivities of approximately 5×10^{-4} S/cm, much higher than those of the ionic conductors alone from which they are derived [MAT 09].

More recently, solid composite electrolytes based on $xLiBH_4$–Li_3N (x = 1, 2, 4) have been studied in order to improve the electrochemical stability of the Li_3N ion conductor. A conductivity of approximately 1×10^{-5} S/cm was obtained with an electrochemical stability up to 3V (vs. Li) [ZHA 17].

2.2.7. *Phosphorus and lithium oxynitride or LiPON*

LiPON with the formula $Li_{3-x}PO_{4-x}N_x$ (with x~0.2) can be presented as a γ-Li_3PO_4 with a lithium deficiency [BAT 92]. This material, which has a conductivity of 3×10^{-6} S/cm at 298 K and an electrochemical stability of between 0 and 5.4 V, is a solid electrolyte essential in thin film microbatteries [YU 97]. Studies on $Li_{3,2}PO_3N$, obtained by cathode sputtering, showed that the incorporation of nitrogen into the vitreous structure allows for a significant increase in conductivity [FLE 11b]. This property has also allowed this material to become the reference electrolyte for thin film batteries. Nevertheless, Solder-Reflow, a method of welding the elements in microelectronics which involves a heat treatment up to 260 °C, induces a modification of the conduction properties of LiPON, which is detrimental since one considers a micro-battery to be just like any other component [FLE 12]. A new vitreous electrolyte based on lithium borophosphate, obtained by cathode sputtering, has been proposed in order to overcome this problem [FLE 13].

2.2.8. *Anti-perovskite lithium-rich solid electrolytes*

In 2012, Zhao *et al.* [ZHA 12] and Braga *et al.* [BRA 13] presented a new class of solid electrolytes, the lithium-rich anti-perovskites (LiRAP), with ionic conductivities σ greater than 10^{-3}S/cm and activation energies close to 0.2–0.3eV.

Zhao *et al.* prepared and characterized the phases Li_3OCl, Li_3OBr, $Li_3OCl_{0.5}Br_{0.5}$ and $Li_3OCl_{0.2}Br_{0.8}$. They are obtained by manual grinding of LiBr (or LiCl) and LiOH. For example, the synthesis of Li_3OCl is carried out according to the reaction:

$$LiCl + 2\,LiOH \rightarrow Li_3OCl + H_2O.$$

The mixture is placed in a quartz tube and heated at 330–360 °C (above the melting temperature, 282 °C for Li_3OCl) under a vacuum for several days, with the water formed during the reaction being removed by a condensate trap connected to a vacuum pump. The molten mixture is slowly cooled or subsequently undergoes quenching to the desired final morphology. The final phase is very hygroscopic and should be stored away from atmospheric humidity.

These compounds are only three examples of this family of lithium-rich (LiRAP) anti-perovskites of the formula Li_3OA with A=F, Cl, Br or I. Different substitutions are possible which can make it possible to obtain a large number of new compounds.

The first is the substitution of a halogen A with another halogen A′ according to the reaction (Figure 2.21(a)):

$$(1-z)\ LiA + z\ LiA' + 2\ LiOH \rightarrow Li_3OA_{1-z}A'_z + H_2O.$$

The phases $Li_3OCl_{0.5}Br_{0.5}$ and $Li_3OCl_{0.2}Br_{0.8}$ were prepared and characterized.

The substitution of Li^+ by a bivalent metal M (such as Mg^{2+}) leading to cationic doping according to the following reaction (Figure 2.21(b)):

$$(1-x)\ LiA + x/2\ MA_2 + 2\ LiOH \rightarrow Li_{3-x}M_{x/2}OA_{1-x} + H_2O.$$

It is possible to create a sub-stoichiometry of LiA and introduce gaps at the A sites and the lithium sites according to the following reaction (Figure 2.21(c)):

$$(1-\delta)\ LiA + 2\ LiOH \rightarrow Li_{3-\delta}OA_{1-\delta} + H_2O.$$

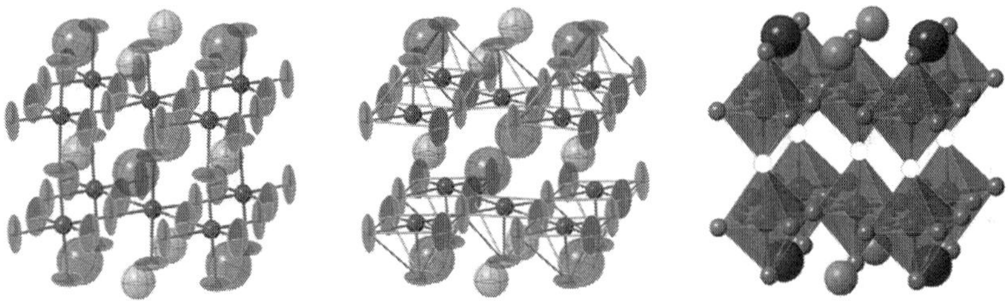

$Li_3OA_{1-z}A'_z$ a) $Li_{3-x}M_{x/2}OA_{1-z}A'_z$ b) $Li_{3-\delta}O(A_{1-z}A'_z)_{1-\delta}$ c)

Figure 2.21. *Anti-perovskite structures illustrating the effects of mixing (a), doping (b) and depletion (c) [151]. For a color version of this figure, see www.iste.co.uk/viallet/batteries.zip*

The anti-perovskites Li_3OCl and $Li_3OCl_{0.5}Br_{0.5}$ have ionic conductivities equal to 0.85×10^{-3} and 1.94×10^{-3} S/cm at room temperature and activation energies, respectively, equal to 0.26 and 0.18 eV (Figure 2.22) [ZHA 12].

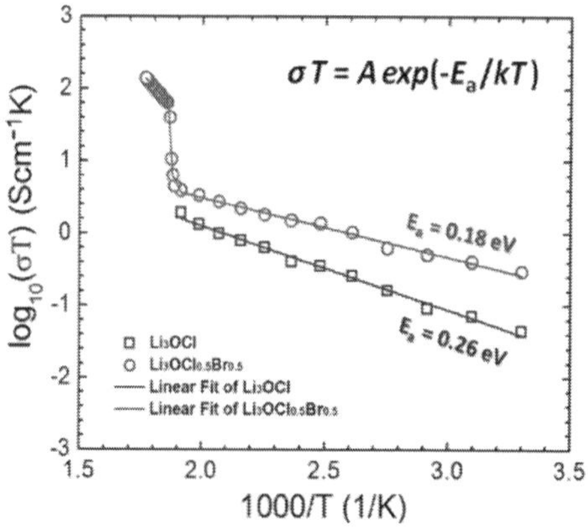

Figure 2.22. *Variation in ionic conductivity with temperature in Arrhenius coordinates, log(σT) vs. 1/T for anti-perovskites Li_3OCl and $Li_3OCl_{0.5}Br_{0.5}$ [ZHA 12]. For a color version of this figure, see www.iste.co.uk/viallet/batteries.zip*

Braga *et al.* [BRA 13] prepared and characterized the anti-perovskites $Li_{3-2x0.04}Ca_{0.04}OCl$, $Li_{3-2x0.005}M_{0.005}OCl$ for the first time in addition to Li_3OCl. Their study shows that doping with a bivalent cation effectively increases the ionic conductivity (Figure 2.23).

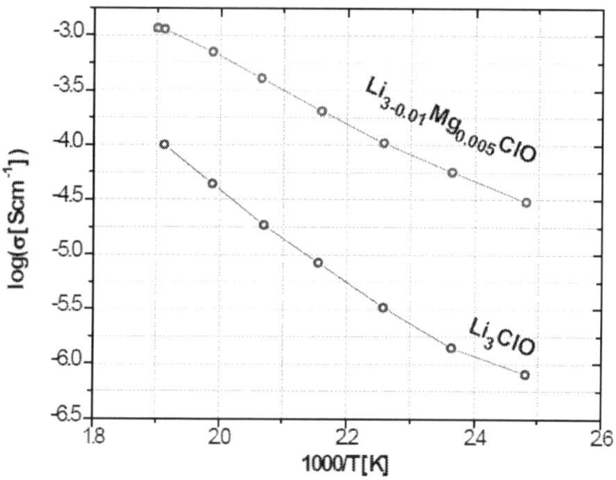

Figure 2.23. *Experimental ionic conductivity measured by impedance spectroscopy for pure and Mg^{2+} doped anti-perovskite Li_3OCl [BRA 13]*

These two teams then combined their efforts in the study of this new family of solid electrolytes, demonstrating the possibility of preparing these LiRAP at high temperature and high pressure [ZHA 14].

Braga *et al.* developed new vitreous solid electrolytes from these anti-perovskite phases based on Li_3OCl [BRA 14] and $A_{3-x}H_xOCl$ (A = Li or Na and $0 < x < 1$) [BRA 16].

2.2.9. *Solid polymer electrolytes*

Several studies on solid-state batteries are devoted to the use of polymer electrolytes. This development began with conductivities of polyethylene oxide (PEO) of alkali metal complexes Wright *et al.* [WRI 75]. In 1978, Armand proposed to use a polymer electrolyte (also known as Solid Polymer Electrolyte – SPE) based on molten POE (polyethylene oxide) containing lithium salts for lithium metal anode batteries [ARM 78, ARM 79]. These polymers based on PEO have been widely used since [DIA 00, GRA 91].

Polymer electrolytes have several advantages, such as increased safety and stability during discharge [BUL 02, MAC 99, DIA 00]. The list of the various polymers, their repeat pattern and their conductivity at room temperature is presented in Table 2.2 from the overview of all-solid-state batteries by Kim *et al.* [KIM 15].

Polymer system	Repeating unit	Electrolyte used	Room temp. conductivity (S cm^{-1})
Polyvinyl alcohol [PVAc]	$(C_2H_4O)n$	(PVA_c)-$LiCF_3SO_3$	10^{-9}
Polymethyl-methacrylate [PMMA]	$-[CH_2C(-CH_3)(-COOCH_3)]n-$	PMMA-$LiClO_4$	4×10^{-3}
Polyethylene oxide [PEO]	$-[CH_2CH_2O]n-$	PEO-$LiBF_4$	10^{-6}
Polyethylene oxide [PEO]	$-[CH_2CH_2O]n-$	PEO_8-$LiClO_4$	10^{-8}
Polyoxymethylene [POM]	$-[CH_2O]n-$	POM-$LiClO_4$	10^{-8}
Polypropylene oxide [PPO]	$-[CH_3CH_2CH_2O]n-$	PPO_8-$LiClO_4$	10^{-8}
Polysiloxane/poly-dimethyl-siloxane [DMS]	$-[(CH_3)_2SiO]n-$	DMS-$LiClO_4$	10^{-5}
PEO linked polysiloxane [PGPS]	PEO-$H(OSiH_2)nOH$	PGPS-$LiClO_4$	10^{-4}
Polypropylene oxide [PPO]	$-[CH_3CH_2CH_2O]n-$	PPO-$LiCl_4$-LiBr-$AlCl_3$	2×10^{-2}

Table 2.2. *Polymer-based solid electrolytes used in all-solid-state batteries*

2.3. Solid sodium-ion conductors

Much more abundant than lithium, sodium is an interesting alternative to lithium-ion batteries. As mentioned in the introduction to explain the interest in all-solid-state batteries, sodium-ion technology has been explored since the 1960s but was later abandoned in favor of lithium-ion technology and its better performance. Limited lithium resources [PAL 12] and increasing demands, especially in electric vehicles, have revived interest in this technology. The latter could meet the safety requirements, long life and low cost and high power required by mass storage systems for intermittent renewable energies, wind or solar. We have also seen that, for the time being, they use sodium-sulfur batteries operating only at high temperatures, at approximately 300 °C, which decreases their energy efficiency.

Following the discovery of beta-alumina ($NaAl_{11}O_{17}$) 80 years ago [BEE 37], sodium ionic conductors are essentially comprised of NASICON [GUI 15], glass and sulfide glass-ceramics, Na_2S–GeS_2, Na_2S–SiS_2, Na_2S-P_2S_5 or even Na_2S–GeS_2–GaS_3 [VAN 07, RIB 80, YAO 08] and the crystallized compound $NaPS_4$. The ionic conductivities of these materials range from 10^{-7} to 10^{-4} S/cm at 25°C, as shown in Figure 2.24. We will only describe the conductors used as solid electrolytes.

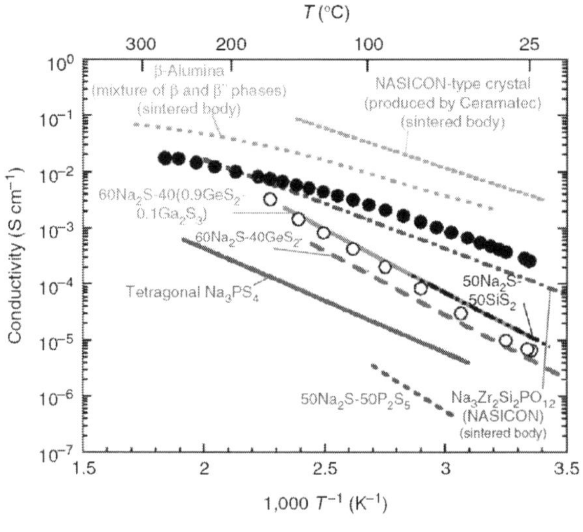

Figure 2.24. *Ionic conductivities of sodium conductive ceramics and glass-ceramics. (○) Glass $75Na_2S$–$25P_2S_5$. (•) Na_3PS_4 glass-ceramic treated at 270 °C [HAY 12]. For a color version of this figure, see www.iste.co.uk/viallet/batteries.zip*

2.3.1. *NASICON compounds*

NASICON or sodium superionic conductor (NASICON) is an extremely well-researched class of materials.

A recent study, based on the collection of existing ionic conductors, of approximately 150 compositions of the general formula $Na_{1+2w+x-y+z}M_w^{(II)}M_x^{(III)}M_y^{(II)}M_{2-w-x-yw}^{(II)}(SiO_4)_z(PO_4)_{3-z}$, was carried out in order to establish correlations between the crystal structure and the ionic conductivity [167].

Depending on the composition, the ionic conductivity varies by several orders of magnitude, from 10^{-13} S/cm for $NaGe_{0.5}Ti_{1.5}(PO_4)_3$ to 10^{-3} S/cm for $Na_{3,2}Hf_2(SiO_4)_{2,2}(PO_4)_{0,8}$.

The most studied NASICON is $Na_3Zr_2(SiO_4)_2(PO_4)$, with x = 2 in the solid solution $Na_{1+x}Zr_2P_{3-x}Si_xO_{12}$ ($0 \leq x \leq 3$), and with a conductivity of 6.70×10^{-4} S/cm at room temperature and 0.2 S/cm at 300 °C [HON 76a, HON 76b, IGN 05]. It is interesting to note that the materials at each end of this binary compound, $NaZr_2(PO_4)_3$ and $Na_4Zr_2(SiO_4)$, are poor conductors.

In 2016, Song *et al.* reported the discovery of a new sodium superionic conductor with the NASICON structure, $Na_{3.1}Zr_{1.95}Mg_{0.05}Si_2PO_{12}$ with a high ionic conductivity at room temperature equal to 3.5×10^{-3} S/cm [SON 16]. The electrochemical stability was also characterized (Figure 2.25). The cyclic voltammetry curve of an Au/$Na_{3,1}Zr_{1,95}Mg_{0,05}Si_2PO_{12}$/Na cell recorded between –0.2 and 9 V at 1 mV/s (Figure 2.25(a)) shows that the ionic conductor has a large electrochemical stability window and no redox activity other than that of sodium below 1 V being detected. The galvanostatic polarization curve of a Na/Na3.1Zr1.95Mg0.05Si2PO12/Na cell with a current density of ± 44 μA/cm2 (Figure 2.25(b)). The polarization potential is stable and equal to 2.7 mV, indicating that the cell is perfectly adapted to the dissolutions and successive depositions of sodium.

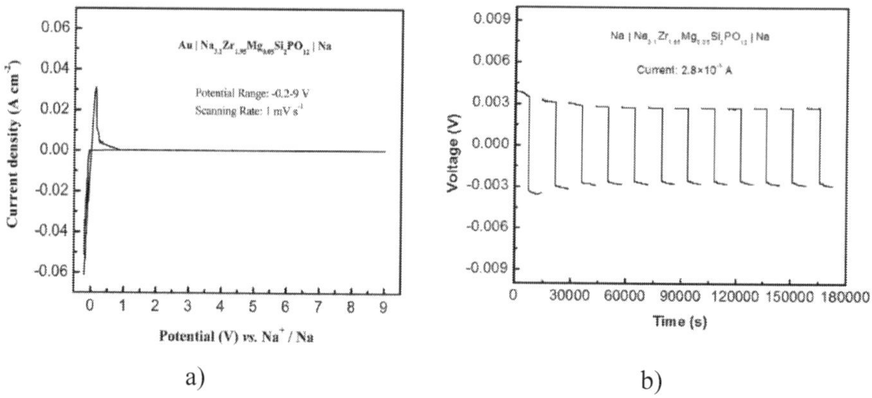

Figure 2.25. *a) The cyclic voltammetry curve of an Au/$Na_{3.1}Zr_{1.95}Mg_{0.05}Si_2PO_{12}$/Na cell recorded between –0.2 and 9 V at 1 mV/s. b) Galvanostatic polarization of a Na/ $Na_{3.1}Zr_{1.95}Mg_{0.05}Si_2PO_{12}$/Na cell under a current density of ± 44 μA/cm²*

We will see later that the conductors $Na_3Zr_2(SiO_4)_2(PO_4)$ (also denoted $Na_3Zr_2Si_2PO_{12}$) and $Na_{3.1}Zr_{1.95}Mg_{0.05}Si_2PO_{12}$ have been incorporated successfully into all-solid-state batteries.

2.3.2. *Na_3PS_4*

In 2012, Tatsumisago and Hayashi in Osaka replicated their mastery of the $Li_2S–P_2S_5$ system on the $Na_2S–P_2S_5$, and more specifically to the compound Na_3PS_4 [HAY 12].

The glass Na_3PS_4 was prepared by mechanosynthesis using a planetary ball mill. The sample obtained after grinding reagents is not crystallized via X-ray diffraction (XRD, Figure 2.26(a)) with a glass transition at 180 °C determined by Differential Thermal Analysis (DTA, Figure 2.26(d)). The only band observed at 420 cm^{-1} on the Raman spectrum of the glass (Figure 45(e)) can be attributed to ortho-thiophosphate ions (PS_4^{3-}), confirming that the composition of the glass is nominal. As shown in Figure 2.26(b), a crystalline phase precipitates after heating the glass Na_3PS_4 at 270 °C, above the first crystallization temperature determined by Differential Thermal Analysis (DTA). The peaks of the glass-ceramic phase obtained are indexed in a cubic cell derived from the quadratic cell of Na_3PS_4 (JCPDS ° 081-1472).

The cubic phase Na_3PS_4 is therefore considered to be the high-temperature phase [JAN 92]. This quadratic phase is effectively equal to that obtained by Hayashi *et al.* by heating to a temperature of 420 °C (diffractogram Figure 2.26(c)) [HAY 12].

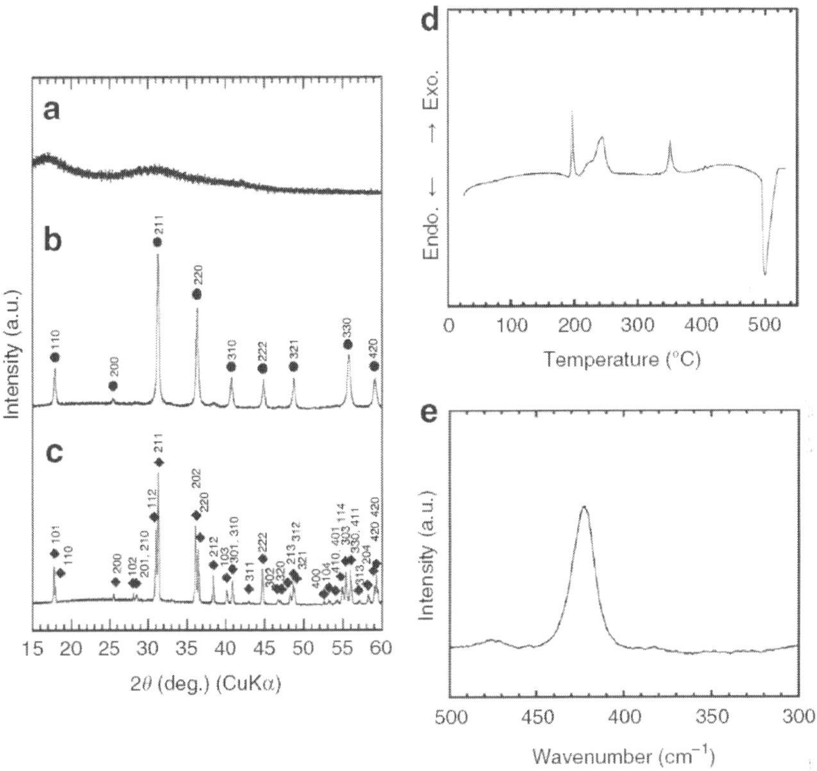

Figure 2.26. *Characterization of the vitreous and glass-ceramic Na_3PS_4 electrolytes. X-ray diffraction patterns of the glass Na_3PS_4 (a), of the glass-ceramic sample heated at 270 °C (b), and the glass-ceramic sample heated to 420 °C (c). The circles and filled diamonds represent the diffraction peaks indicating the cubic phase and the quadratic phase, respectively (JCPDS ° 081-1472). (d) DTA curve of the glass. (e) Raman spectrum of the glass*

The stable form of sodium ortho-thiophosphate Na_3PS_4 at low temperatures is normally the quadratic form, as shown in Figure 2.28, a structure organized around PS_4^{3-} tetrahedra. At high temperatures, the narrowing of these tetrahedra leads to a phase transition toward a cubic symmetry with a higher conductivity. Hayashi *et al.* obtained this form directly by heat treatment of the glass $75Na_2S–25P_2S_5$ at 270 °C, undoubtedly due to planetary grinding. The cubic glass-ceramic Na_3PS_4 has a conductivity of 2×10^{-4} S/cm at ambient temperature, making it the best ion conductor in its class. This cubic phase appears to be metastable, since annealing the glass at a temperature of 420 °C results in the crystallization of the quadratic phase, reducing the conductivity to 1×10^{-6} S/cm.

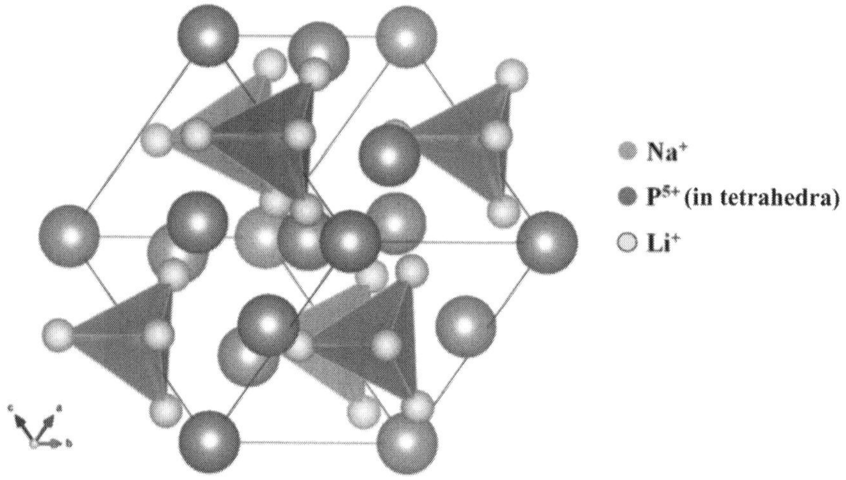

Figure 2.27. *Crystallographic structure of Na_3PS_4. Space group: P -42$_1$c, a=6.9520(4) Å and c = 7.0757(5) Å [JAN 92]. For a color version of this figure, see www.iste.co.uk/viallet/batteries.zip*

This instability can be observed on the ionic conductivity curve as a function of the temperature shown in Figure 2.24 which, from 150 °C, exhibits slope discontinuity suggesting a modification of the conduction properties, which can be initiated by a cubic–quadratic phase transition.

The characterization of the electrochemical stability window of this highly conductive cubic form of Na_3PS_4 is shown in Figure 2.28. The successive dissolving and deposition currents of sodium were observed at approximately 0 V versus Na^+/ Na, and no significant oxidation current was observed up to 5 V versus Na^+/Na. This experiment suggests that the electrolyte has a wide electrochemical window of 5 V versus Na^+/Na and is electrochemically stable versus metallic sodium.

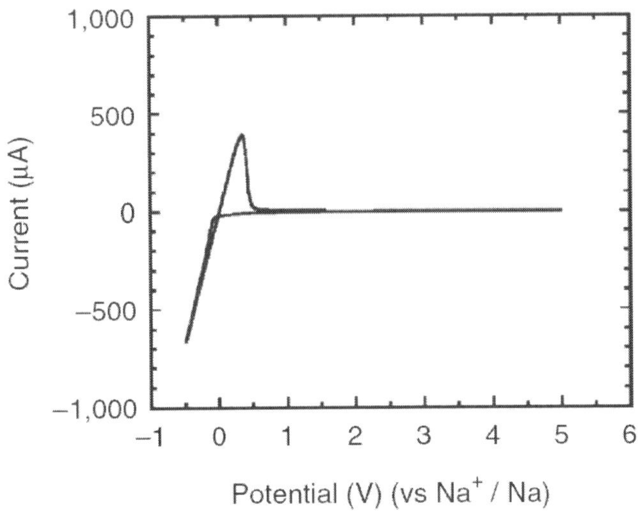

Figure 2.28. *Electrochemical stability at 25 °C of the glass-ceramic electrolyte Na_3PS_4, prepared at 270 °C. Cyclic voltammogram trace of a cell consisting of a stainless steel disc of the glass-ceramic and a sodium sheet (scanning potential of 5 mV/s)*

By using the crystalline Na_2S precursor with a higher purity (99.1 %), the cubic glass-ceramic phase Na_3PS_4 is obtained directly by grinding for the composition $75Na_2S–25P_2S_5$ (mol%) for 1.5 hr of grinding [HAY 14]. A heat treatment at 270 °C of the glass obtained improves the crystallinity and a conductivity of 4.6×10^{-4} S/cm is obtained, twice as high as that obtained from another precursor, as described earlier. The conductivity could be increased to 7.4×10^{-4} S/cm by adding 6% of Na_4SiS_4 as carried out by the team in Osaka [TAN 14]. Glass-ceramics with a composition of (100–x) Na_3PS_4–xNa_4SiS_4 (mol%) with $0 \leq x \leq 10$ were prepared by mechanical grinding followed by heat treatment, all with conductivities greater than 10^{-4} S/cm at ambient temperature. Conductivity increases concomitantly with the increasing Na_4SiS_4 content for $0 \leq x \leq 6$. Glass-ceramics with the composition $25 \leq x \leq 100$ have conductivities of 10^{-7} to 10^{-5} S/cm.

Recently, the same team reported the synthesis of Na_3PS_4 from a liquid phase process [YUB 15]. A yellow solution is obtained following the reaction of P_2S_5 and Na_2S in *N*-methylformamide (NMF). The cubic phase Na_3PS_4 precipitates after evaporation of the solvent. The ionic conductivity of the electrolyte obtained via this route is equal to 2.6×10^{-6} S/cm at ambient temperature. This conductivity is not as high as for the phase prepared by planetary grinding followed by heat treatment, but this solution may be useful for coating electrode particles in order to improve the electrochemical performance of all-solid sodium batteries.

3

All-Solid-State Battery Technology Using Solid Sulfide Electrolytes

Many groups have started working on the so-called monolithic all-solid-state batteries, made up of layers of several hundred microns thick in order to pass from μAh/cm² obtained in micro-batteries to mAh/cm² from solid sulfide electrolytes.

3.1. Monolithic Li-ion "all-solid-state" batteries

3.1.1. *The first "all-solid-state" batteries*

In the 1980s, the French groups that initiated work on conductive glass also developed the first "all-solid-state" batteries.

In 1982, Ribes *et al.* reported the development of an all-solid-state battery for the first time in collaboration with the company GIPELEC [RIB 82, CAR 83].

In 1984, Ménétrier reported the first 100% inorganic "all-solid-state" battery, based on the results of Knutz *et al.* with lithium nitride Li_3N [MEN 84, KNU 83]. As seen in Figure 3.1, the battery consists of a stack of a TiS_2 electrode layer, a layer of the Li_2S–B_2S_3–LiI glass electrolyte and a Li–Al alloy sheet used as a counter-electrode. The capacity delivered by the battery can be increased by producing a mixed TiS_2 electrode/electrolyte. This battery already meets two criteria necessary for the realization of a high-performance "all-solid-state" battery using a vitreous and glass-ceramic electrolyte: the use of a mixed electrode and a lithium alloy (this criterion will be developed at a later stage).

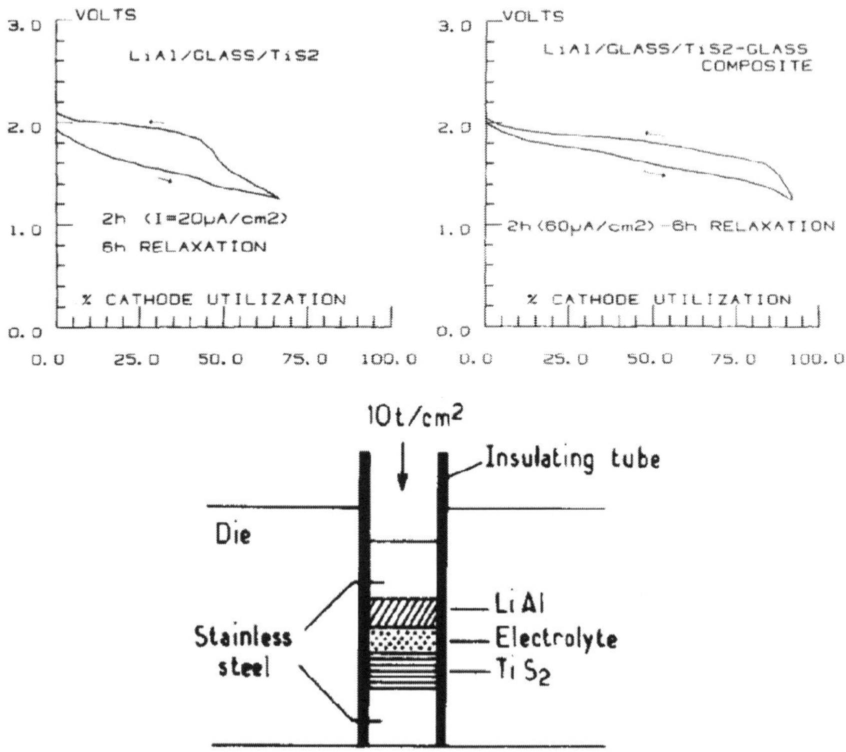

Figure 3.1. *Cycling curves and diagram of the three layers* *$Li–Al/Li_2S–B_2S_3–LiI/TiS_2$ battery [MEN 84]*

3.1.2. *Second generation "all-solid-state" batteries*

In the late 1990s, work by Takada and Tatsumisago and their team resulted in the emergence of a new generation of high-potential "all-solid-state" batteries [TAK 96, TAT 00, MIN 00, KOM 01]. The chosen electrode material is $LiCoO_2$, already marketed in Sony Li-ion batteries.

At that time, protocols for assembling and electrochemical characterization of "all-solid-state" batteries were established and still exist today. The battery is made up of a 10-mm-diameter polycarbonate insulating matrix by cold pressing of the electrode and electrolyte layers (Figure 3.2).

The thickness of the battery obtained is around a millimeter, that is to say 100 times greater than a thin-film battery.

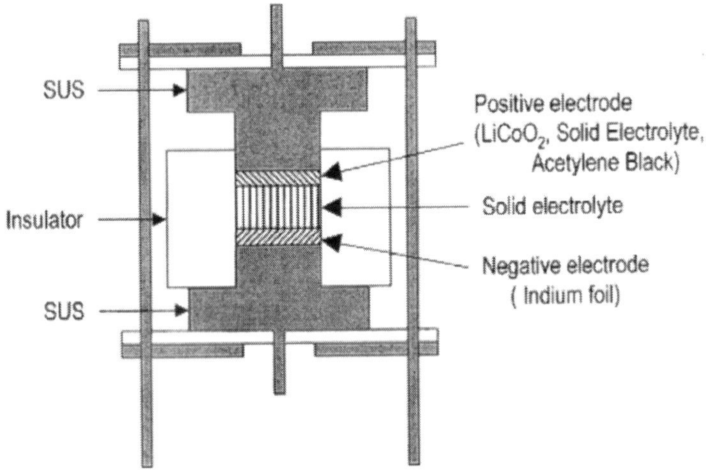

Figure 3.2. *Diagram of an "all-solid-state" battery, according to Komiya et al. [KOM 01]*

Composite electrodes are obtained by mixing the active material with the electrolyte and a conductive additive. The proportions of each component constitute a key parameter that strongly influences the performance of the batteries. The addition of the electrolyte ensures good transport of the lithium ions but also optimizes the interface with the electrode. Figure 3.3 shows the SEM images of the various components of a battery [MIZ 05a], namely the electrode material, here $LiCoO_2$ (a), the electrolyte, in this case the electrolyte $80Li_2S\text{-}20P_2S_5$ (b), and electronic conductive additives, acetylene black (c), vapor grown carbon fibers (VGCF) (d), titanium nitride TiN (e) or even metallic nickel Ni (f).

With the separating electrolyte layer needing to be thick enough to support the shaping and thin enough so as not to cause excessive polarization, its mass was fixed at 80 mg. The counter-electrode used is an indium–lithium alloy. This alloy is obtained spontaneously after 12 hr through contact between a sheet of indium and lithium. Although the electrolytes exhibited good stability with respect to lithium based on electrochemical stability tests, the use of the alloy was preferred, but no explanation was given for this choice [TAK 96]. A set of screws was used in order to maintain constant pressure during cycling, necessary to allow an optimal exchange of the lithium ions at the interfaces.

The cycling tests were carried out at a current density of 64 μA/cm² or one of its multiples at ambient temperature. These values are indeed consistent in articles on "all-solid-state" batteries using a sulfur-based electrolyte.

Figure 3.3. *SEM images of $LiCoO_2$ (a), $80Li_2S–20P_2S_5$ electrolyte (b) and various conductive additives: acetylene black (c), VGCF (d), TiN (e) and Ni (f) [IEO 16]*

These criteria resulted in the assembly of high-performance "all-solid-state" batteries delivering capacities between 70 and 110 mAh/g at ambient temperature [TAK 96, TAT 00, KOM 01, MIZ 02] (Figure 3.4).

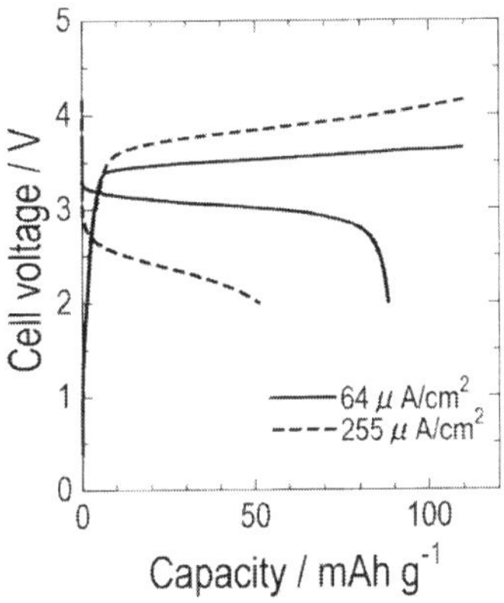

Figure 3.4. *Cyclic curve of a $LiCoO_2/Li_2S–SiS_2–Li_4SiO_4$/In–Li battery [KOM 01].*

With respect to high potential positive electrode materials, very few materials apart from $LiCoO_2$ have been tested in the literature on "all-solid-state" batteries. We thus find the spinels $LiMn_2O_4$ and $LiNi_{0,5}Mn_{0,5}O_4$, which have retention capacities of 60–70 mAh/g [KIT 10a, MIZ 03], as well as lamellar $LiNi_{1/3}Mn_{1/3}Co_{1/3}O_2$, 110 mAh/g [KIT 10b] for around 20 cycles. In addition, since the main argument for "all-solid-state" batteries is the electrochemical stability of the electrolyte, it is surprising not to find studies on materials with very high potentials such as $LiCoPO_4$, which raise many questions about the determination of the electrochemical stability of these electrolytes, questions which we shall try to answer in the next chapter.

With regard to the conductive additive used in electrodes, Mizuno demonstrated that the use of Vapor Grown Carbon Fiber (VGCF) allows for optimum electronic percolation leading to better retention of capacity [MIZ 05a]. In addition, electrodes prepared without a conductive additive but with 70% $LiCoO_2$ showed excellent electrochemical performances, ensuring the transport of electrons [MIZ 05b].

Since 2005 and the discovery of $80Li_2S–20P_2S_5$ glass ceramic, Hayashi and his team studied many negative electrode materials, some of which are summarized in Table 3.1. Most of the studies center around the compounds M_xS_y or M_xP_y, the aim being to create a continuous network based on phosphorus and sulfur between the electrode and the electrolyte. Among these compounds, $Cu_xMo_6S_{8-y}$ has a capacity retention of 100 mAh/g over 2000 cycles [NAG 09].

Material	Current density	Capacity retention	Reference
$SnS–P_2S_5$	64 µA/cm²	400 mAh/g after 70 cycles	[HAY 05]
α-Fe_2O_3	64 µA/cm²	800 mAh/g (first)–350 mAh/g (20th)	[KIT 07]
NiS	1.3 mA/cm²	360 mAh/g over 50 cycles	[NIS 09]
$Cu_xMo_6S_{8-y}$	1.28 mA/cm²	100 mAh/g over 2000 cycles	[NAG 09]
NiP_2	64 µA/cm²	600 mAh/g over four cycles	[HAY 09a]
Black P	3.8 mA/cm²	500 mAh/g over 150 cycles	[NAG 11a]
$SnP_{0.94}$	1.3 mA/cm²	1000 mAh/g	[ASO 10]

Table 3.1. *Examples of negative electrode materials studied in "all-solid-state" batteries*

Nevertheless, the best results are obtained with the compound $Li_4Ti_5O_{12}$, the material of choice for the negative electrode of "all-solid-state" batteries. Indeed, this compound has been greatly studied in the field of Li-ion accumulators because it does not exhibit any volume expansion during the insertion/disinsertion phases [OHZ 95] and can withstand extremely high cycling regimes [KIT 09]. Thus, a capacity retention of 140 mAh/g over 700 cycles could be obtained for a battery using the $Li_2S–P_2S_5–P_2S_3$ electrolyte [MIN 09, TAT 08, MIN 11]. This result was obtained at 100 °C and at a current density of 12.7 mA/cm², that is to say a regime of 1 Li^+ per hour (Figure 3.5).

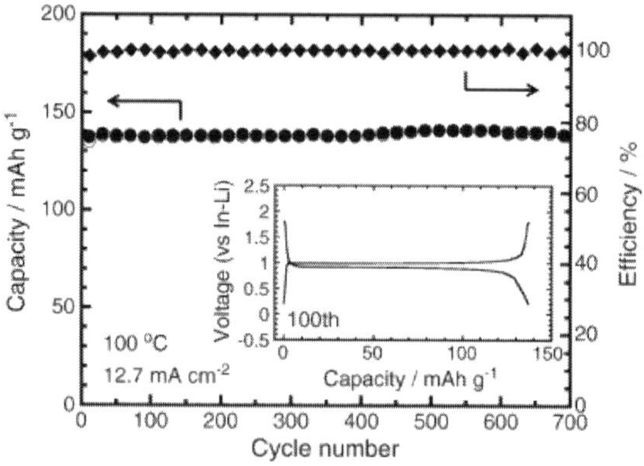

Figure 3.5. *Cycling performance of an all-solid-state In–Li/70Li_2S·27P_2S_5·1P_2S_3 glass-ceramic/$Li_4Ti_5O_{12}$ battery with a current density of 12.7 mA/cm² at 100 °C. The inset shows the charge–discharge curves of the cell at the hundredth cycle [MIN 11]*

In the literature, all "all-solid-state" batteries using a sulfur-based electrolyte were tested with an indium-based or indium–lithium-based counter electrode, although the electrolytes proved to be stable against lithium metal. It was not until 2012 that Nagao presented the reactivity between the electrolyte $80Li_2S–20P_2S_5$ and lithium [NAG 12a] through the galvanostatic cycling study of a symmetrical $Li/80Li_2S–20P_2S_5/Li$ battery. The potential difference induced by the passage of a current in the cell causes the transfer of the lithium from one electrode to the other by dissolution and then deposition. As seen in Figure 3.6(a), this potential difference suddenly drops after about 10 hr, and the battery is completely short-circuited after 30 h.

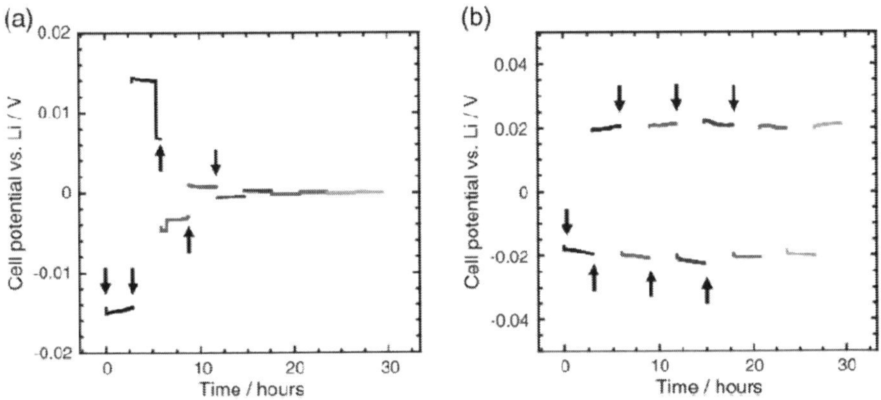

Figure 3.6. *a) Galvanostatic cycling curve of a $Li/80Li_2S–20P_2S_5/Li$ battery and b) Li/Li-PVD/80Li2S–20P2S5/Li-PVD/Li [NAG 12a]. For a color version of this figure, see www.iste.co.uk/viallet/batteries.zip*

This phenomenon could be limited by the deposition of a thin layer of lithium by Physical Vapor Deposition (PVD), allowing for a lithium/ electrolyte contact surface, and therefore, a better reversibility of the dissolution/deposition as shown in Figure 3.7(b).

3.1.3. *Toward High Performance Batteries*

The performances of the batteries mentioned earlier are essentially due to the fact that the sulfur-based glass and glass-ceramics can be easily deformed, making it possible to obtain excellent contacts between the particles of active material and the electrolyte during shaping. The interfaces are therefore key to "all-solid-state" battery performances, and their optimization will enable the assembly of batteries with high power densities.

In 2006, Ohta *et al.* demonstrated the formation of an insulating layer on the surface of $LiCoO_2$ during cycling [OHT 06]. The hypothesis put forward by this team is that the significant difference in ionic conductivity between $LiCoO_2$ and the solid electrolyte causes a migration of Li^+ ions toward the electrode material, leading to an ionically depleted, and therefore highly insulating, interface. This is referred to as a space charge layer.

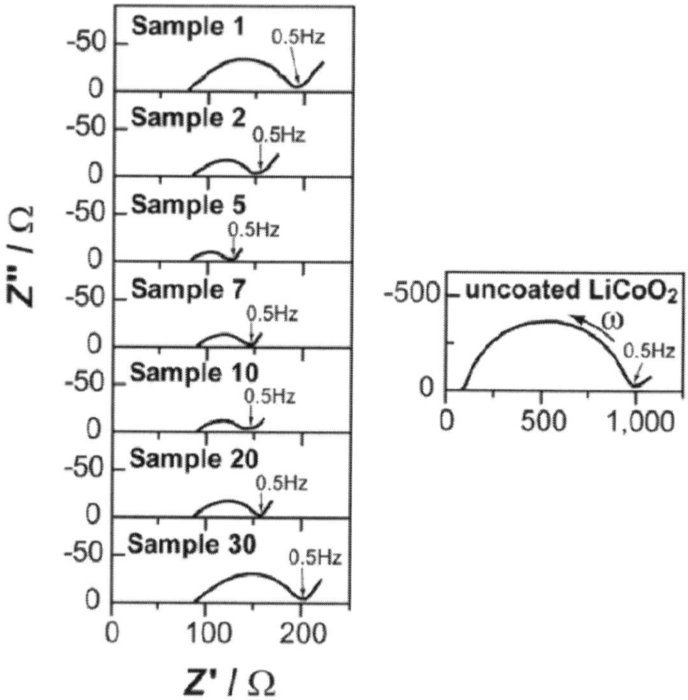

Figure 3.7. *Representation of the resistance of the space charge layer as a function of the coating thickness by $Li_4Ti_5O_{12}$. The sample number corresponds to the thickness in nm [OHT 06]*

To limit this phenomenon, $LiCoO_2$ is coated with a nanoscale layer of $Li_4Ti_5O_{12}$ which serves as a buffer. A coating layer of only 5 nm of $Li_4Ti_5O_{12}$, obtained by sol-gel reaction, makes it possible to reduce the resistance by a factor of 20, from 1000 to only 50 Ω (Figure 3.7). This resistance is then reduced to about 10 Ω per coating for an 8–12 nm layer of $LiNbO_3$, a more conductive material than $Li_4Ti_5O_{12}$ [OHT 07]. This space charge layer has also been observed for other positive electrode materials such as $LiMn_2O_4$ and $LiNi_{1/3}Mn_{1/3}Co_{1/3}O_2$. Coating with $Li_4Ti_5O_{12}$ has improved electrochemical performance [KIT 10, KIT 11a].

In 2009, Sakuda *et al.* also studied this space charge layer by choosing to coat the $LiCoO_2$ particles with a glass layer of Li_2SiO_3 [SAK 08, SAK 09b], a material with excellent ionic conductivity and with no transition metal capable of reacting. Transmission Electron Microscopy analysis of the interface between $LiCoO_2$ and the $Li_2S–P_2S_5$ electrolyte revealed that lithium ion migration was not the only phenomenon leading to the formation of an insulating layer [SAK 09a].

As shown in Figure 3.8(i), an interdiffusion of Co, P and S elements takes place between the material and the electrolyte, forming an insulating interfacial layer of approximately 100 nm thickness. The coating of the $LiCoO_2$ particles with Li_2SiO_3 makes it possible to reduce the thickness of this layer to about 20 nm (Figure 3.8(ii)).

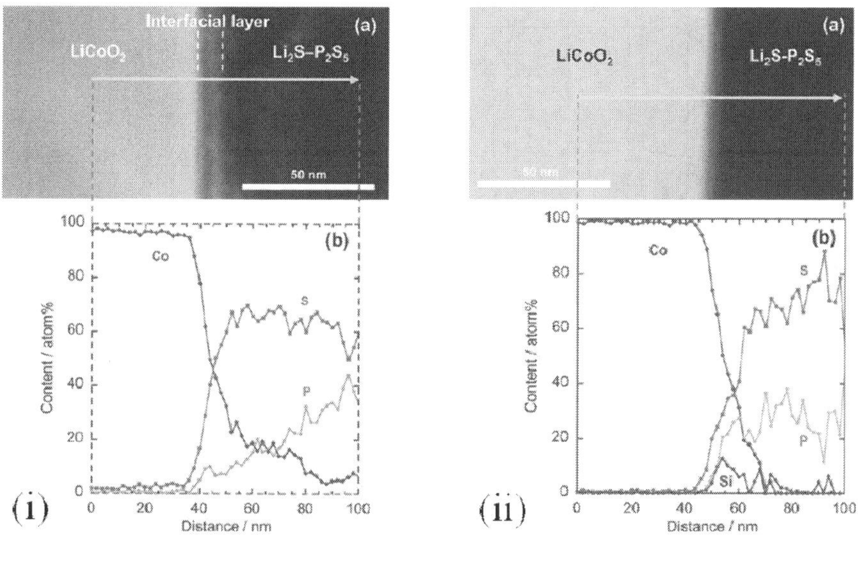

Figure 3.8. *HAADF-STEM image and EDX profile of the interface between $LiCoO_2$ and the $Li_2S–P_2S_5$ electrolyte after an initial charging without coating (i) and with Li_2SiO_3 coating (ii) [SAK 09a]. For a color version of this figure, see www.iste.co.uk/viallet/batteries.zip*

Coating $LiCoO_2$ with Li_2SiO_3 made it possible to assemble "all-solid-state" batteries with very high power densities. Thus, a Li_2SiO_3–$LiCoO_2$/Li_2S–P_2S_5/In–Li cell delivered a capacity of 100 mAh/g at 100 °C and a current density of 40 mA/cm², i.e. a regime of 20C (1 Li^+ in 3 min). This battery was also tested at –30 °C and delivered a capacity of 60 mAh/g at a regime of C/40 (1 Li^+ in 40 hr).

Sakuda *et al.* also carried out double-layer coating tests on $LiCoO_2$ [SAK 09c, SAK 10, SAK 11]. In order to decrease the proportion of electrolyte in the electrode, and thus to improve the specific energy as illustrated in Figure 3.9, $LiCoO_2$, previously coated with $LiNbO_3$, is again coated with a layer of Li_2S–P_2S_5 by Pulsed Laser Deposition (PLD) [SAK 11]. This second layer allowed a battery containing 90 wt% $LiCoO_2$ to deliver a capacity of 100 mAh/g at a regime of C/20 (1 Li^+ in 20 hr).

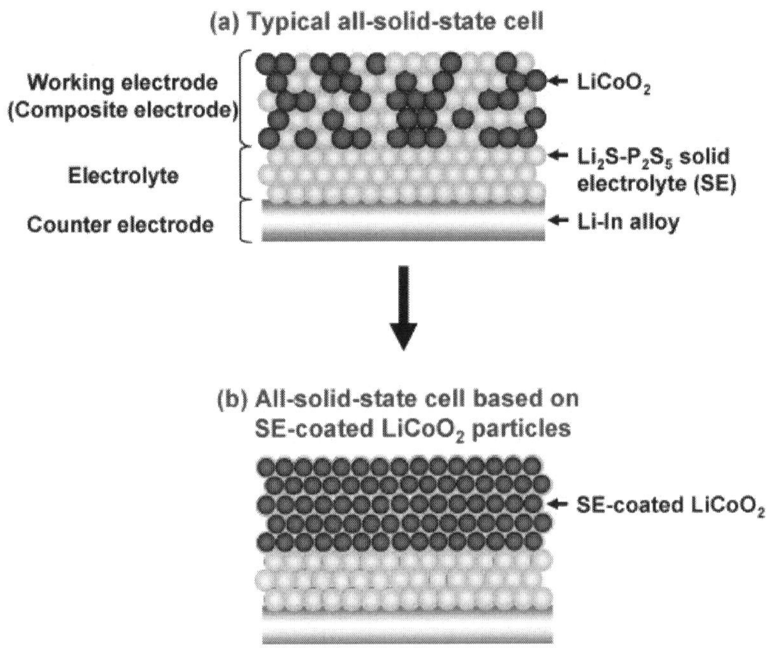

Figure 3.9. *Diagram of an "all-solid" battery using $LiCoO_2$ coated with an oxide (a) and $LiCoO_2$ coated with a Li_2S–P_2S_5 electrolyte (b). For a color version of this figure, see www.iste.co.uk/viallet/batteries.zip*

In 2011, Hayashi *et al.* experimented with a new way of shaping the "all-solid-state" batteries that allowed a significant improvement in the electrochemical performances: Hot Pressing [KIT 11b]. This technique involves heat treating a powder while maintaining a high pressure in order to improve its compactness. This method resulted in an increase in the conductivity of the electrolyte by increasing its compactness while improving the electrode/electrolyte interfaces by a slight melting of the electrolyte. The first complete "all-solid-state" $LiCoO_2/Li_2S–P_2S_5/Li_4Ti_5O_{12}$ battery was assembled, delivering a capacity of 110 mAh/g of $LiCoO_2$.

In 2013, this same team reviewed various techniques used to improve the interfaces between the solid electrolyte and the electrode materials [TAT 13] as shown in Figure 3.10.

The first approach, presented in the top right, consists of preparing nanocomposites, i.e. composites in which the materials are at the nanometric scale, using mechanical grinding [HAY 08a, NAG 12b]. A Li_2S-based composite as the active material, $Li_2S–P_2S_5$ as a solid electrolyte (SE) and acetylene black (AB) (an electronic conductor additive) was prepared by high-energy planetary mill grinding. STEM-EEL analyses showed that the Li_2S particles were 200 nm and that of AB 100 nm. The all-solid-state batteries using the nanocomposite showed a higher charge and discharge capacity and better cyclability than when the three components Li_2S, SE and AB were prepared by manually grinding.

The second approach, shown in the upper right-hand corner, already presented earlier, consists of coating the active material with a thin layer of SE electrolyte. $LiCoO_2$ particles previously coated with a thin layer of $LiNbO_3$ were coated with a layer of $Li_2S–P_2S_5$ with a conductivity of 10^{-4} S/cm by pulsed laser deposition [SAK 11].

The third technique is the use of a vitreous electrolyte in a supercooled liquid state. The mixture of vitreous electrolyte, $Li_2S–P_2S_5$, and active material, $LiCoO_2$, is heated to a temperature close to the glass transition temperature, 210 °C in the case of $Li_2S–P_2S_5$. After cooling to room temperature, the electrolyte–electrode contacts are maximized. The images obtained by Scanning Electron Microscopy show that the grain boundaries and the pores in the particles of $LiCoO_2$ are filled with vitreous domains and the battery assembled with such a composite electrode showed a greater reversible capacity.

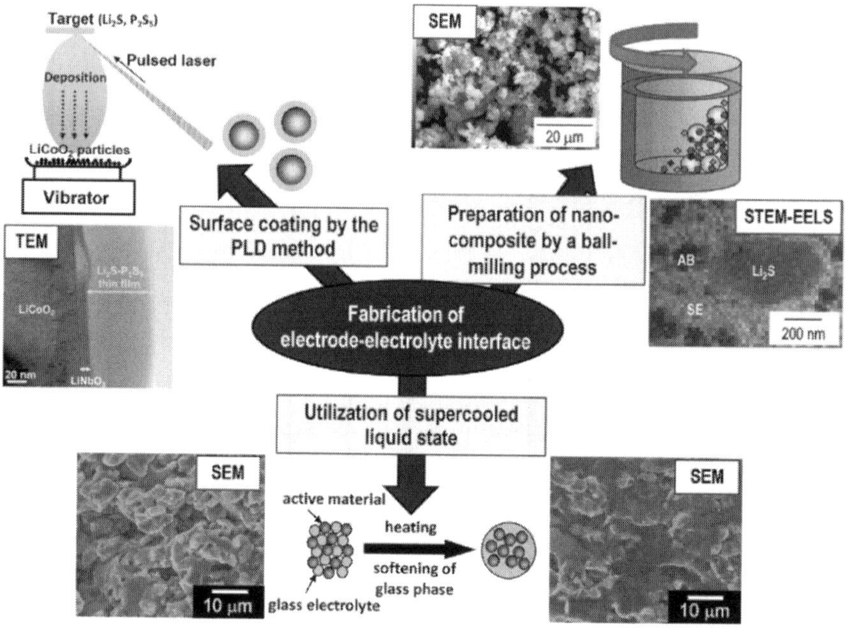

Figure 3.10. *Diagram of the different approaches used to form optimized electrolyte/electrode interfaces [TAT 13]. For a color version of this figure, see www.iste.co.uk/viallet/batteries.zip*

3.1.4. *Batteries using lithium argyrodite electrolytes*

3.1.4.1. *$Li_4Ti_5O_{12}$, first electrode material used*

The first "all-solid-state" battery was presented by Stadler in 2010, in collaboration with Deiseroth, and using Li_6PS_5Br as the solid electrolyte [STA 10]. The composite electrode of this cell consists of a 70/30 (wt%) mixture of $Li_4Ti_5O_{12}$ coated in carbon (c-LTO) and an electrolyte. The counter-electrode used is a Li–Al alloy (0.4 V vs. Li^+/Li^0) (Figure 3.11(a)). Extremely low electrochemical performances, 10 mAh/g for c-LTO at C/10, were obtained at room temperature, due to the ionic conductivity of the electrolyte, 1×10^{-4} S/cm. The capacity delivered by the battery could be greatly improved by increasing the cycling temperature (130 mAh/g for c-LTO at 75 °C, Figure 3.11(b)). Nevertheless, the cell exhibits poor capacity retention due to the volume expansion of the Li–Al counter-electrode during cycling.

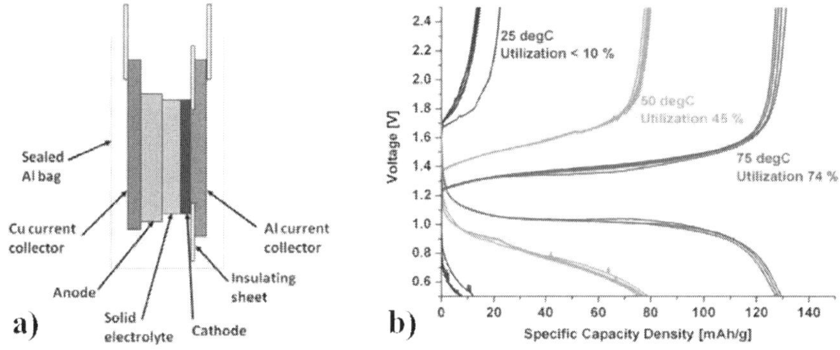

Figure 3.11. *a) Diagram of the c-LTO/Li_6PS_5Br/Li–Al cell used for the cycling tests, the diameter is 11 mm for a thickness of approximately 200 µm. b) Galvanostatic cycling curve of the cell at 25, 50 and 75 °C, at 200 µA/cm², i.e. a rate of C/10 [DEI 08]. For a color version of this figure, see www.iste.co.uk/viallet/batteries.zip*

The second "all-solid-state" battery was proposed by Rao using Li_6PS_5Cl as the solid electrolyte, $Li_4Ti_5O_{12}$ as the active material and carbon black as the conductive additive [DEI 08]. The planetary milling of the constituent elements of the composite electrode made it possible to considerably increase the capacity delivered by the battery, compared with a conventional mixture made with mortar (Figure 3.12). A $Li_4Ti_5O_{12}$/Li_6PS_5Cl/Li cell has a relatively good capacity retention of 60 mAh/g after 60 cycles at room temperature. This is the second reported example of an "all-solid-state" battery using lithium metal as the counter electrode.

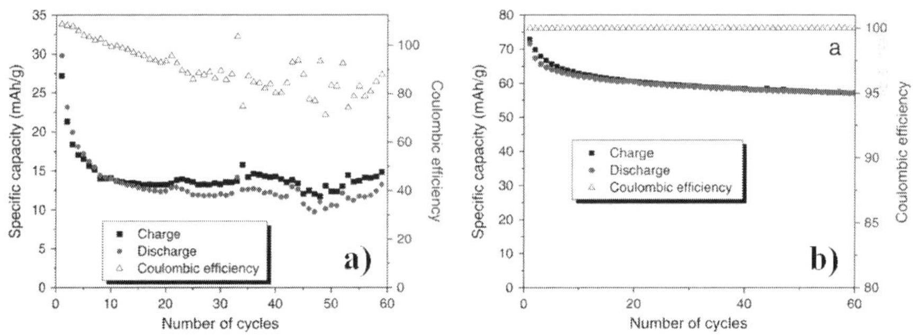

Figure 3.12. *Capacity retention curves for a $Li_4Ti_5O_{12}$/Li_6PS_5Cl/Li cell whose composite electrode has been prepared by grinding in mortar (a) and planetary grinding (b) [RAO 11]. For a color version of this figure, see www.iste.co.uk/viallet/batteries.zip*

Research into the incorporation of argyrodite in solid-state batteries by Boulineau *et al.* has highlighted several major technological challenges inherent to "all-solid-state" batteries [BOU 13a, BOU 13b]. These technological limits are related to the nature of the counter electrode, the composition and the mass of the electrodes, as well as that of the electrolyte, the four main points analyzed in this study.

With regard to the nature of the counter electrode, lithium metal is the ideal candidate but its use is limited to low current densities, and in particular, used with the positive electrode material $LiCoO_2$, with indium giving better results with the latter. $Li_4Ti_5O_{12}$ is interesting as a negative electrode. We have also seen that composite electrodes have at least 30 wt% of electrolyte [HAY 03, MIN 10], which implies a considerable "dead" volume in the battery. Since the addition of the electrolyte takes place to the detriment of the active material and results in a reduction in the capacitance delivered by the composite and thus in a reduction in the specific energy of the battery, it is necessary to optimize the composition in order to limit the electrolyte proportion as much as possible.

In the literature, "all-solid-state" batteries are conventionally assembled with 80 mg of electrolyte and 10 mg of composite electrode [MIN 10]. Also given that 30% of the mass of the composite electrode is also electrolyte, this means that 92% of the mass of the powder used to make an "all-solid-state" battery is electrolyte that does not intervene in the electrochemical activity. It is therefore necessary to minimize the thickness of the electrolyte layer without modifying its mechanical or conduction properties.

The mass of the electrodes is also a challenge related to the two previous points. Indeed, it is possible to improve the specific energy of the battery by increasing the composite electrode mass. Considering the previous example, using 100 mg of electrode instead of 10 mg reduces the dead weight to 61% of the total mass of the battery. The study showed that increasing the electrode mass, and therefore its thickness, causes a problem with lithium ion diffusion throughout the electrode. However, an electrode twice as thick does not deliver twice the capacity and it is necessary to bring the battery to 100 °C to optimize its performance.

Composite $LiCoO_2$/Arg and composite $Li_4Ti_5O_{12}$/Arg half-cells were first assembled and then the compositions of the composite electrodes were optimized. Capacity retention on the first cycles of the two best-performing composites, 54%LCO/36%Arg-Cl and 36%LTO/54%Arg-Cl, are shown in Figure 3.13.

The LCO composite exhibits a high irreversibility of about 20% in the first cycle and a capacity retention of 45 mAh/g, which gradually decreases (Figure 3.13(a)). In addition, the battery has a polarization of approximately 150 mV which increases as cycling progresses, which may explain the decrease in capacity. Nevertheless, this composite exhibits a Coulombic efficiency of 98% in the third cycle.

The LTO composite also has a very high irreversibility, about 25%, during the first charge (Figure 3.13(b)) and an excellent capacity retention of 26 mAh/g during the first 10 cycles. The Coulombic efficiency is 99% after the cycle, with a polarization of only 50 mV, increasing very slightly during cycling.

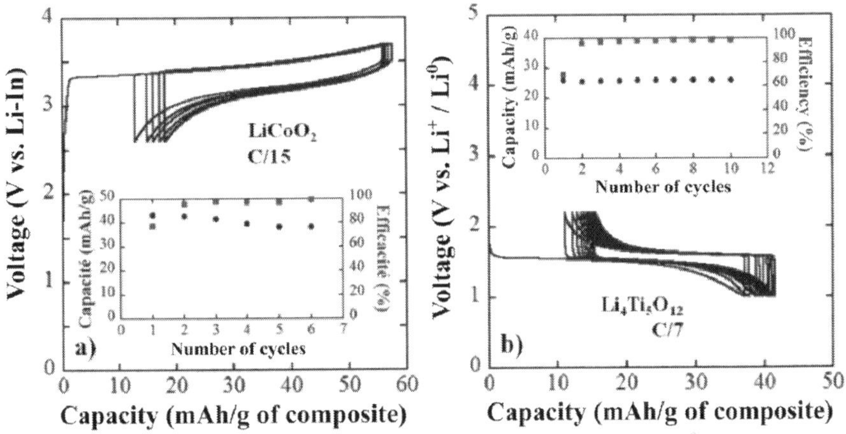

Figure 3.13. *Charge/discharge curves for the composites 54%LCO/36%Arg-Cl/In (a) and 36%LTO/54%Arg-Cl/Li (b). The insets show Coulombic capacity and efficiency according to the number of cycles. For a color version of this figure, see www.iste.co.uk/viallet/batteries.zip*

3.1.4.2. *$Li_4Ti_5O_{12}$//Argyrodite//$LiCoO_2$ full-cell*

On the basis of these results, a full-cell using $LiCoO_2$ as the positive electrode and $Li_4Ti_5O_{12}$ as the negative electrode was assembled [BOU 13a, BOU 13b]. To evaluate the correct shaping of these batteries, a section was observed by scanning electron microscopy, the images are shown in Figure 3.14. The first thing that can be noticed is that the cold press assembly leads to very dense batteries (Figure 3.14(a)). Taking into account the mass and the density of each of the battery components, we have evaluated the compactness of the electrolyte at ~90% and that of the electrode at more than 86%. As for the electrolyte, we find the same compactness as when measuring the ionic conductivity on pellets of Li_6PS_5Cl alone. For the composite electrode, which we will see again later, the compactness is dependent on its composition and will be one of the key parameters for the realization of compact "all-solid-state" batteries.

The second notable point is the good interface, since it is relatively difficult to differentiate between the electrolyte (A) and the composite electrode (B), suggesting a weak interface and therefore favorable to good cyclability. When the interface between the electrode and the electrolyte is magnified (Figure 3.14(b)), there is no well-defined differentiation between the electrolyte contained in the composite electrode and that acting as the separator.

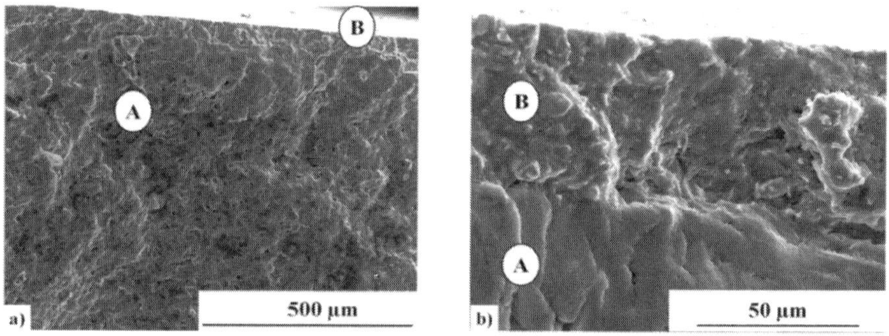

Figure 3.14. *Scanning Electron Microscopy images of the edge of an "all-solid-state" battery assembled by cold pressing. General image (a) and magnification of the electrode/electrolyte interface (b)*

This excellent shaping by simple cold pressing is, once again, ensured by the glass-ceramic nature of the electrolyte, and therefore, by the high deformability of its amorphous part. The charge/dischare curve of this "rocking-chair" battery is shown in Figure 3.15.

On the basis of the results obtained for the half-cells, a complete LCO/Arg-Cl/LTO cell was assembled and characterized.

The capacity delivered by the electrode is only 26 mAh/g, compared to 45 mAh/g observed in the half-cell (Figure 3.15). With the capacity of the full-cell being of the same order of magnitude as that of the LTO half-cell, the performance of the battery appears to be limited by the negative electrode.

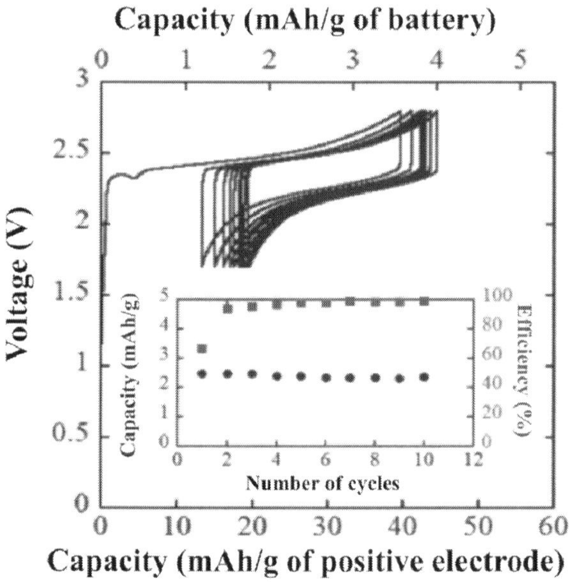

Figure 3.15. *Charge/discharge curve of the complete LCO/Arg-Cl/LTO battery, at 64 μA/cm² and at ambient temperature. For a color version of this figure, see www.iste.co.uk/viallet/batteries.zip*

Doubling the negative electrode mass, and therefore its thickness, does not result in an increase in capacity, as expected, since the latter does not deliver more capacity than in the half-cell. This phenomenon can be explained by poor diffusion of lithium in the thickness of the electrode, which can be improved by an increase in temperature (as we shall see later).

In order to evaluate the specific energy of the complete battery, the capacity was expressed as a function of the total mass. The battery provides a capacity of 2 mAh/g with a retention profile identical to that of the LCO half-cell and a coulombic efficiency of 98 % (Figure 3.15). The voltage difference of 2.45 V makes it possible to estimate the specific energy of this battery at 4.9 Wh/kg, which is 10 times lower than those of a lead battery.

In the literature, a single "all-solid-state" battery, and more specifically a $LiCoO_2/Li_4Ti_5O_{12}$ battery, was reported by Hayashi *et al.* [MIN 10], all others using indium or indium–lithium alloy. The first charge–discharge curves and electrochemical performances at 64 μA/cm² of a $Li_4Ti_5O_{12}/80Li_2S–20P_2S_5$ glass/$LiNbO_3–LiCoO_2$ all-solid-state battery prepared by hot-pressing (— and B) and cold pressing (– C) are shown in Figure 3.16.

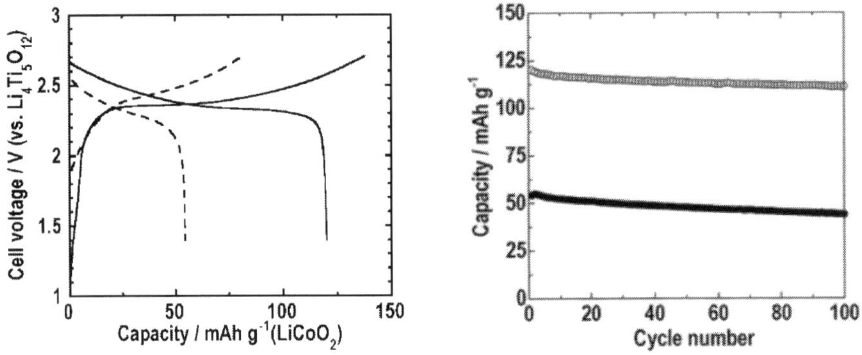

a) b)

Figure 3.16. *(a) First charge–discharge curves and (b) electrochemical performances at 64 μA/cm² of an all-solid-state $Li_4Ti_5O_{12}/80Li_2S–20P_2S_5$ glass/$LiNbO_3–LiCoO_2$ prepared by hot-pressing (— and B) and cold pressing (– C)*

The temperature stability of the LCO-Arg-Cl and LTO/Arg-Cl composites was evaluated by Differential Scanning Calorimetry (DSC) and ThermoGravimetric Analysis (TGA) before testing the batteries at 100 and 150 °C. No reactivity between electrode material and electrolyte was detected. This stability was also evaluated for $LiCoO_2$ by *in situ* X-ray diffraction at different temperatures. The different diffractograms obtained between 25 and 125 °C showed no structural modification of the two phases in the presence of $LiCoO_2$ and Arg-Cl, and no growth of a parasitic phase resulting from a chemical reactivity between these two compounds. This study showed that the Arg-Cl/$LiCoO_2$ and Arg-Cl/$Li_4Ti_5O_{12}$ mixtures are stable and can be used to evaluate the performance of a battery using Argyrodite at high temperatures.

The temperature cycling tests were carried out in an LRCS-developed cell shown in Figure 3.17. This new cell is composed of two Swagelok® bodies on either side of a 13-mm-diameter silica tube.

(a) (b)

Figure 3.17. *a) Diagram of the cell used for temperature cycling and b) Photograph of the furnace into which it is introduced*

The battery prepared in the form of a pellet by cold pressing is introduced into the tube and two steel plungers make electrical contact. Pressure is maintained within the cell by means of a spring. The cell is sealed by eight Teflon joints distributed on each side of each Swagelok® body. The cell is then introduced into a specially developed furnace. The Swagelok® bodies are located outside the furnace so that the Teflon seals do not melt and the spring can continue to provide contact pressure. The dimensions of the cell are such that the optimum zone of heating is located exactly where the pellet is in order to be sure of the temperature at which the batteries are cycling.

Figure 3.18 shows the capacity retention curve of a LCO/Arg-Cl/LTO full-cell using the composites 38% LCO (or LTO)/57% Arg-Cl/5% VGCF. The cell has a capacity of 7.4 mAh/g at a C/10 rate and a specific energy of 18.2 Wh/kg. The increase in temperature makes it possible to greatly reduce the polarization of the cell when the speed is increased, due to the increase in the ionic and electronic conductivity of each of the battery components. Thus, while the capacity delivered by the battery dropped drastically to C/5 at ambient temperature, no decrease was observed at 100 °C. As we can see, the battery has excellent power resistance and a capacity of 6.2 mAh/g, i.e. 15 Wh/kg at a 2C rate.

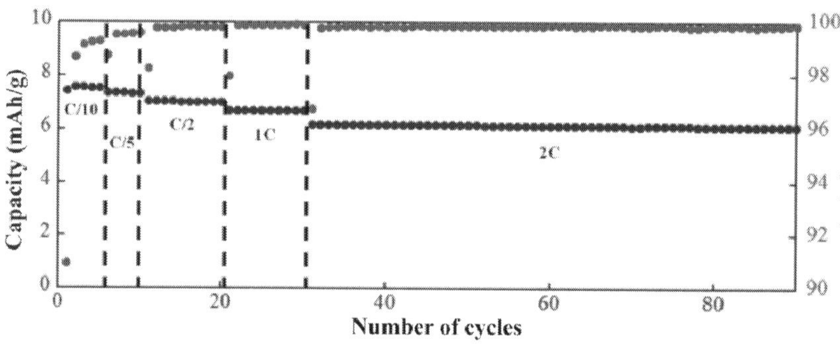

Figure 3.18. *Capacity retention curve of a LCO/Arg-Cl/LTO battery assembled with 20 mg of electrode, tested at 100 °C at different cycling rates. For a color version of this figure, see www.iste.co.uk/viallet/batteries.zip*

3.1.5. $Li_{10}XP_2S_{12}$ (X = Ge, Si, Sn) phase in the structure LGPS

In 2011, Kanno presented the first cycling curves of a cold-pressed $LiCoO_2/Li_{10}GeP_2S_{12}$/In cell with the Ge-based electrolyte of the LGPS structure. This battery has good cyclability with a theoretical capacity reaching 120 mAh/g, under a current density of 14 Ma/g. However, only eight cycles are shown. With the data provided, it is not possible to determine the cycling rate (Figure 3.19) [KAM 11].

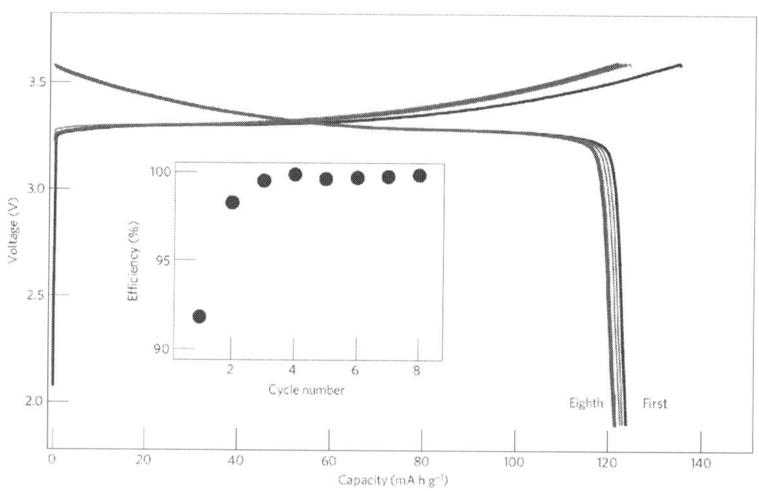

Figure 3.19. *Cycling curves of a $LiCoO_2/Li_{10}GeP_2S_{12}$/In cell [KAM 11]. For a color version of this figure, see www.iste.co.uk/viallet/batteries.zip*

In 2012, Kato at Toyota, in partnership with Kanno *et al.*, compared the performance of $LiCoO_2/Li_{10}GeP_2S_{12}/Li_4Ti_5O_{12}$ (Figure 3.20(a)) and $LiCoO_2/P_2S_5–Li_2S/Li_4Ti_5O_{12}$ cells (Figure 3.20(b)). Performances at high rates are improved for the cell using $Li_{10}GeP_2S_{12}$, with 40 mA h/g discharge capacity at a current density of 57.5 Ma/cm^2 (the second cell cannot cycle at this rate). However, no assessment of cycling performance was reported [KAT 12].

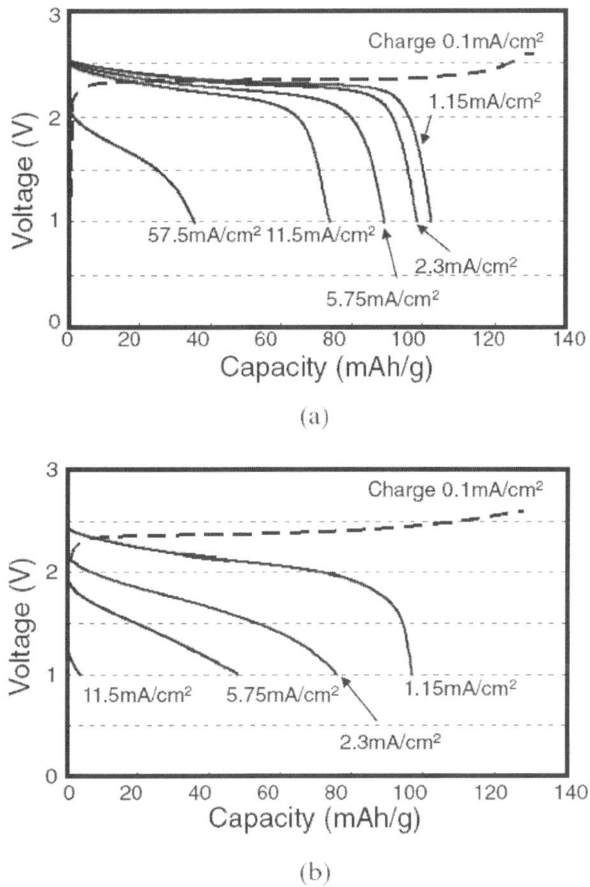

Figure 3.20. *Charge–discharge curve of an all-solid-state battery using (a) $Li_{10}GeP_2S_{12}$ and (b) $75Li_2S{\cdot}25P_2S_5$.*

In 2014, Shin *et al.* assessed the performance of several TiS_2/electrolyte/Li–In cells, using various combinations of Li_3PS_4 (LPS,

1.0 × 10^{-3} S/cm) and $Li_{10}GeP_2S_{12}$ (LGPS, 6.2 × 10^{-3} S/cm) for the composite positive electrode and the electrolyte (Figure 3.21) [SHI 14]:

– composite (TiS_2–LPS)/LPS electrolyte/Li–In;

– composite (TiS_2–LGPS)/LGPS electrolyte/Li–In;

– composite (TiS_2–LGPS)/LPS electrolyte/Li–In;

– composite (TiS_2–LPS)/bi-layer LGPS–LPS electrolyte/Li–In;

– composite (TiS_2–LGPS)/bi-layer LGPS–LPS electrolyte/Li–In.

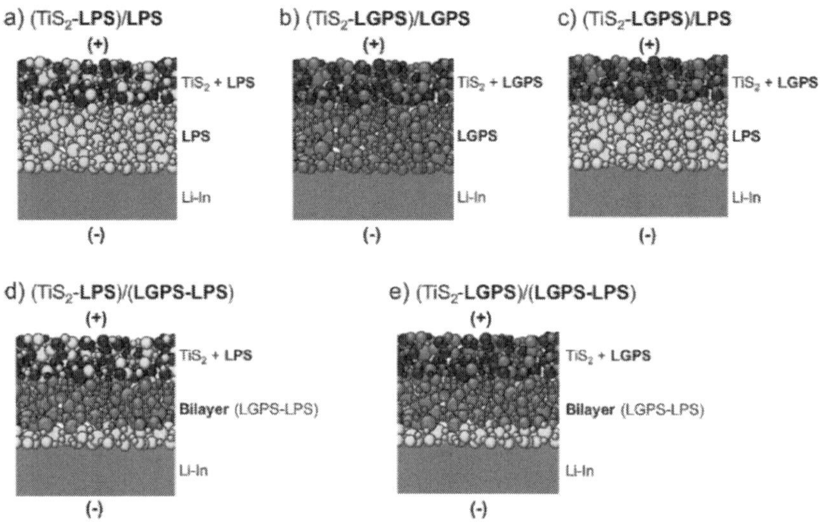

Figure 3.21. *Different configurations of all-solid-state TiS_2/electrolyte/Li–In cells using different combinations of composite and solid electrolyte layers. For a color version of this figure, see www.iste.co.uk/viallet/batteries.zip*

This study shows that the configuration with a layer of Li_3PS_4 between the negative electrode and the $Li_{10}GeP_2S_{12}$ electrolyte makes it possible to obtain the best performances with 60 mAh/g (blue curve) at 20 °C in a potential range between 1.5 and 3 V versus Li^+/Li at 30 °C (Figure 3.22).

These better performances are due to the fact that $Li_{10}GeP_2S_{12}$ is not very stable at low voltages and that a protective layer between the negative electrode and the electrolyte is necessary in order to limit the parasitic reactions. In addition, the low conductivity of Li_3PS_4 does not allow for good performances at high rates if it is used as an electrolyte alone or within the composite with TiS_2 [SHI 14].

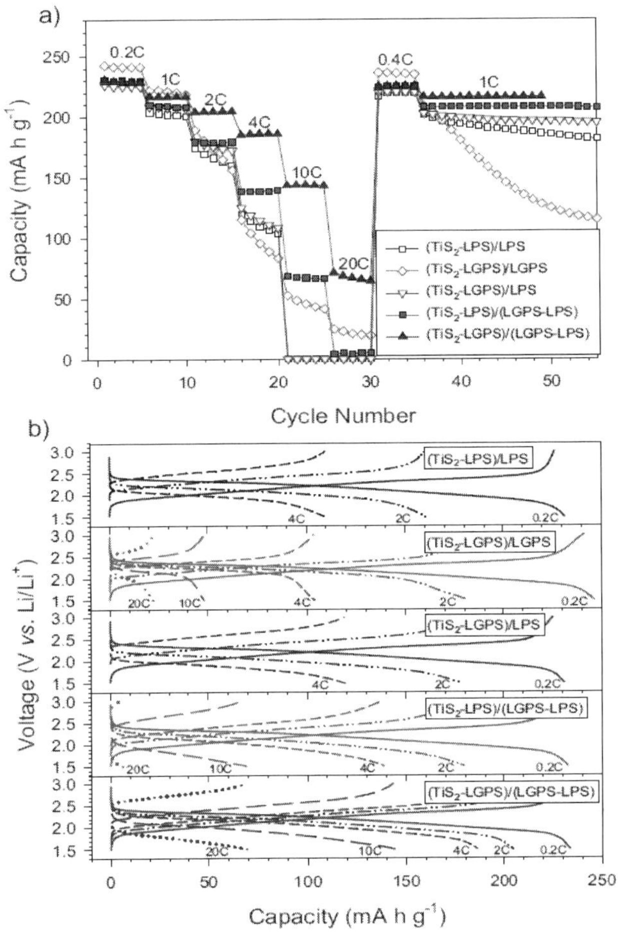

Figure 3.22. *a) Variation in charge capacity as a function of the number of cycles and the rate. b) Charge–discharge curves of the various all-solid-state batteries, between 1.5 and 3 V versus Li^+/Li at 30 °C. For a color version of this figure, see www.iste.co.uk/viallet/batteries.zip*

In 2015, Yin *et al.* compared the performances of $LiNi_{0.8}Co_{0.15}Al_{0.05}O_2$(NCA)/electrolyte/Li–In cells using $Li_{10}GeP_2S_{12}$ and a thio-LiSICON $Li_{3.25}Ge_{0.25}P_{0.75}S_4$ as electrolytes, with the respective conductivities 8.27×10^{-3} and 2.03×10^{-3} S/cm (Figure 3.23) [YIN 15]. The performance of the cell using $Li_{10}GeP_2S_{12}$ was better even at high rates (1C) due to its higher ionic conductivity. These improved cycling performances are also due to a lower interface resistance between NCA and $Li_{10}GeP_2S_{12}$.

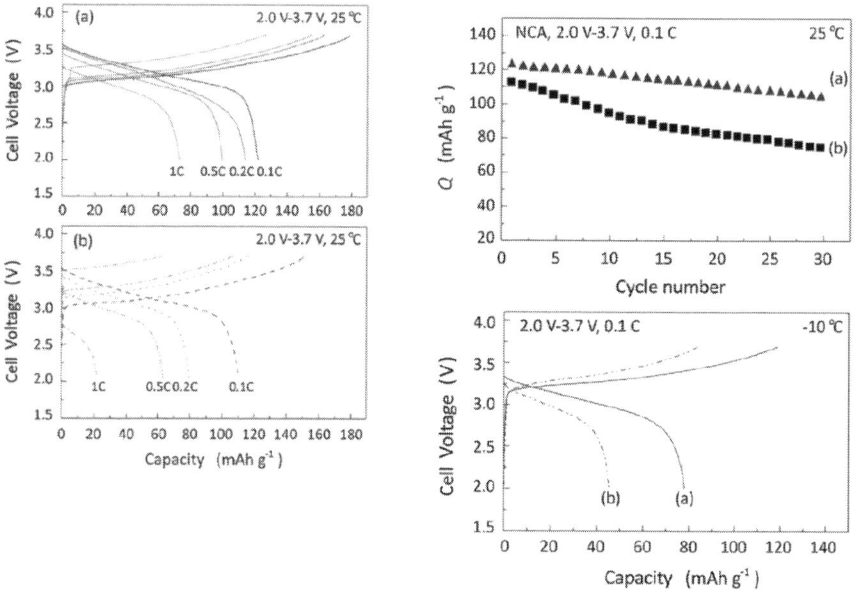

Figure 3.23. *a) Initial charge–discharge curve for NCA/$Li_{10}GeP_2S_{12}$/Li–In and b) NCA/$Li_{3.25}Ge_{0.25}P_{0.75}S_4$/Li–In; top right, cycling performances at 25 °C and (bottom right) initial charge–discharge cycles for NCA/$Li_{10}GeP_2S_{12}$/Li–In (a), and NCA/$Li_{3.25}Ge_{0.25}P_{0.75}S_4$/Li–In (b) at 0.1 °C between 2.0 and 3.7 V at −10 °C [YIN 15]. For a color version of this figure, see www.iste.co.uk/viallet/batteries.zip*

Whiteley *et al.* also compared the performance of $LiNi_{1/3}Mn_{1/3}Co_{1/3}O_2$/electrolyte/Li cells with $Li_{10}GeP_2S_{12}$ (LGPS) and $Li_{10}SiP_2S_{12}$ (LSiPS) as electrolytes. The less expensive silicon-based analog of the Kanno-germanium phase has a high conductivity of 2.3×10^{-3} S/c at room temperature.

For both cells, polarization increases with cycling (Figure 3.24). However, the increase in polarization is much greater in the case of the cell with $Li_{10}GeP_2S_{12}$ (blue curve) than with $Li_{10}SiP_2S_{12}$ (red curve). The phase $Li_{10}SiP_2S_{12}$ would have a better chemical and electrochemical stability with respect to lithium and would be more compatible with the high-potential cathode materials than the phase $Li_{10}GeP_2S_{12}$.

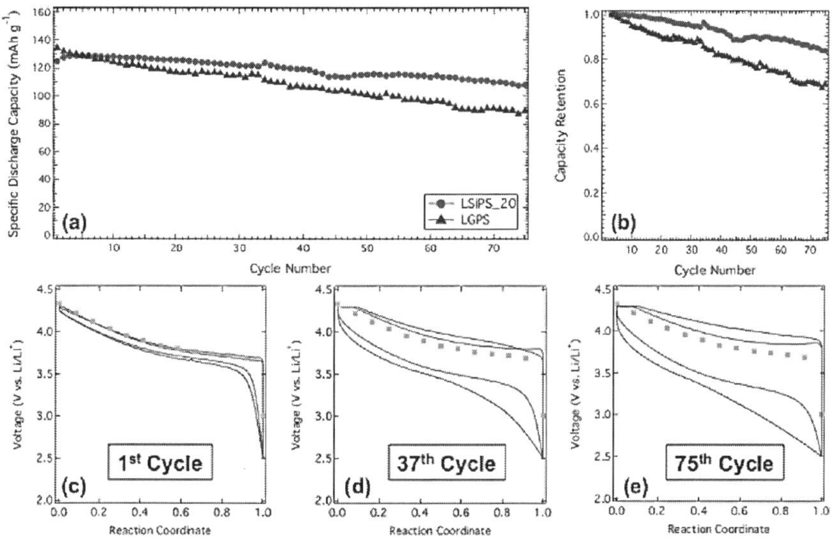

Figure 3.24. *Charge capacity of the first 75 cycles of $Li(Ni_{1/3}Mn_{1/3}Co_{1/3})O_2$ electrodes with LSiPS (red) or LGPS (blue) at C/10, b) capacity retention of cells shown in a) normalized with the third cycle, c) first discharge and second charge, d) 37th discharge and 38th charge, e) 75th discharge and 76th charge. The profiles of c–e are normalized at 100% of the capacity for a better comparison. The orange points have the equilibrium values obtained using GITT performed on $Li(Ni_{1/3}Mn_{1/3}Co_{1/3})O_2$ in a liquid cell. For a color version of this figure, see www.iste.co.uk/viallet/ batteries.zip*

Thus, different batteries produced reveal the instability of the $Li_{10}GeP_2S_{12}$ phase, both at low and at high voltages. This instability has, however, given rise to an interesting concept, a battery made from a single material, presented by Han *et al.* [HAN 15]. $Li_{10}GeP_2S_{12}$ effectively exhibits a high conductivity of 10^{-2} S/cm and its electrochemical stability window is reported up to 5 V versus Li^+/Li. This large window is due to the small oxidation and reduction kinetics due to its low electronic conductivity ($\sigma_e \approx 10^{-9}$ S/cm) because the upper limit of the intrinsic thermodynamic stability window is calculated at 2.5 V [MO 11]. These same calculations show that LGPS oxidizes into P_2S_5, S and GeS_2 at high voltages and is reduced to Li_2S, Li_3P and $Li_{15}Ge_4$ at low voltages. Sulfur, the oxidation product of LGPS, and $Li_{15}Ge_4$, the reduction product, are well-known positive and negative electrode materials, respectively. Thus, oxidation and reduction reactions could take place beyond the stability limits, the kinetics being accelerated by the addition of carbon and thus by the use of an LGPS-C composite.

Moreover, the Li–S bonds in LGPS are similar to those observed in Li_2S, a positive electrode active material with a theoretical capacity of 1166 mAh/g at ≈ 2.2 V [NAG 12, YAN 12], making use of LGPS as an active material even more likely. GeS_2 is known as a negative electrode active material with a theoretical capacity of 823 mAh/g at a potential close to 0.5 V [KIM 08, CHO 13]. Pure LGPS could serve as a solid electrolyte, and when associated with carbon in an LGPS-C composite, it could serve as both negative and positive electrodes. To validate this concept, an all-solid-state battery was assembled by superimposing an LGPS-C composite cathode, the LGPS electrolyte and a composite LGPS-C anode as shown in Figure 3.25(b).

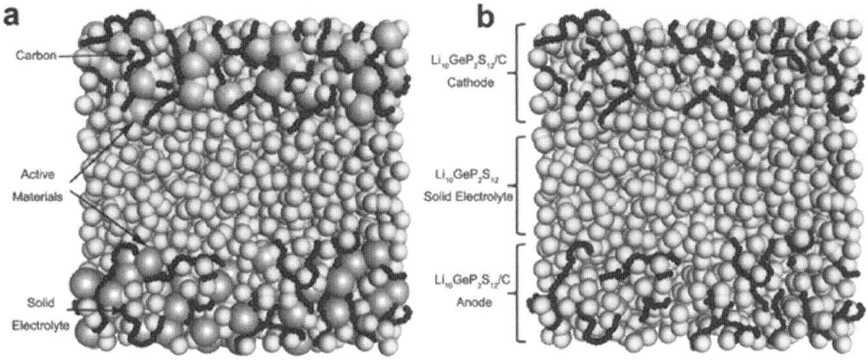

Figure 3.25. *Schematic diagrams of a conventional all-solid-state battery (a) and an all-solid-state battery prepared with LGPS in the electrolyte and at the cathode, and the anode as the active material (b) [HAN 15]. For a color version of this figure, see www.iste.co.uk/viallet/batteries.zip*

The electrochemical activity in the presence of carbon has been demonstrated. The chemical stability window determined from cyclic voltammetry on a Li/LGPS/Pt cell at a rate of 0.05 mV/s is presented in Figure 3.26(i) and shows that LGPS is stable up to a high potential of 5 V, in line with the previously published results. On the other hand, when platinum is replaced by a layer of LGPS-C (75:25 wt% mixture of LGPS-C) to form a Li/LGPS/LGPS-C half-cell (Figure 3.26(ii)), a reversible redox peak at low voltages between 0.0 and 0.5 V is observed when recording between 0.0 and

2.0 V. Another redox phenomenon appears at high voltages between 1.6 and 2.7 V when the recording is between 1.0 and 3.5 V [HAN 15, supporting information]. This result shows that an LGPS-C composite can serve as a positive electrode between 1.6 and 2.7 V and a negative electrode between 0.0 and 0.5 V.

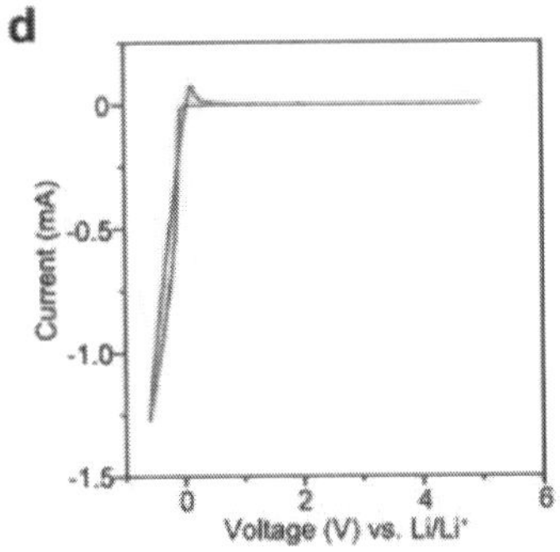

(i)

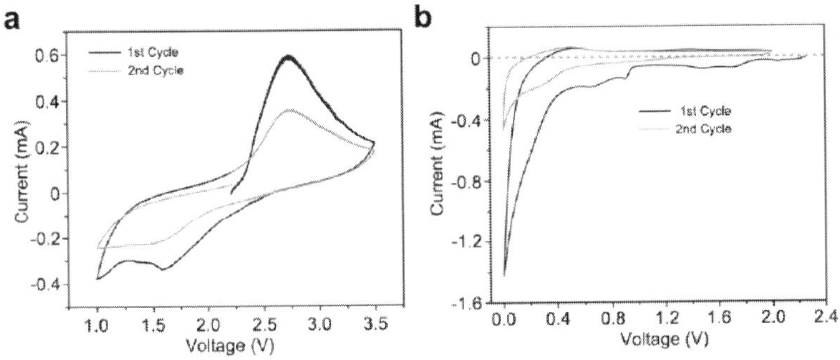

(ii)

Figure 3.26. *(i) Cyclic voltammetry curve of a Li/LGPS/Pt cell in a potential range between –0.6 and 5.0 V (recording rate 0.05 mV/s). (ii) Cyclic voltammetry curves of a Li/LGPS/LGPS-C/Pt cell in the potential ranges 1.0–3.5 V (a) and 0–2 V at a rate of 0.05 mV/s [HAN 15]. For a color version of this figure, see www.iste.co.uk/viallet/batteries.zip*

The electrochemical performances of the LGPS-C electrodes were first tested with a liquid electrolyte in button cells and then in a Swagelok device specially designed for the study. LGPS-C half-cells were characterized in different voltage ranges to evaluate the performance of composite electrodes considered as positive and negative electrodes (Figure 3.27). LGPS-C composite electrodes were obtained by mixing the LGPS and the carbon by mechanical milling for 20 hr to ensure a good distribution of the carbon within the composite. LGPS-C as a positive electrode has a reversible capacity of 267, 140 and 80 mAh/g for current densities of 10 to 50 and 100 mA/g. LGPS-C as a negative electrode has reversible capacities of 130, 60 and 36 mAh/g for the same values.

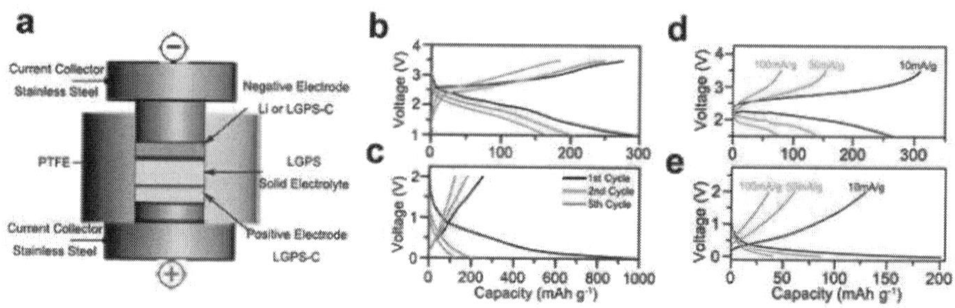

Figure 3.27. *Electrochemical performances of a cathode and a composite anode $Li_{10}GeP_2S_{12}$-C and solid electrolyte $Li_{10}GeP_2S_{12}$. a) Schematic representation of the all-solid-state battery cycling device. Charge/discharge curves of all-solid-state Li/LGPS/LGPS-C cells at current densities 10 mA/g in the potential range 1.5–3.5 V (b) and 0.0–2.0 V (c). Charge–discharge curves of all-solid-state Li/LGPS/LGPS-C cells for different current densities in these same ranges [HAN 15]. For a color version of this figure, see www.iste.co.uk/viallet/batteries.zip*

These results made it possible to assemble an all-solid-state LGPS-C positive/LGPS/LGPS negative cell, the mass ratio of negative LGPS-C to positive LGPS-C was set at 3 to compensate for the capacity difference and the high irreversible capacity of the anode during the first discharge (Figure 3.27(c)). No delamination was observed as shown in Figure 3.28(a) and LGPS is continuously distributed over the entire thickness of the pellet, the positive LGPS-C layers, LGPS electrolyte and negative LGPS-C being, respectively, 222, 330 and 148 μm. The electrochemical response of this monolithic battery at a current density of 10 mA/g between 0.0 and 2.5 V shows that the battery delivers a reversible capacity of 104 mAh/g by LGPS. The cycling performance was evaluated by cycling the battery under different currents, revealing that the reversible capacity decreases from 104 to 70 then 37 mAh/g from 10 to 50 and 100 mA/g, respectively. The reversible capacity increases from 70 to 96 mAh/g by LGPS when the temperature increases from room temperature to 50°C for a current density of 50 mA/g due to the reduction of the electrode-to-electrolyte interface resistances.

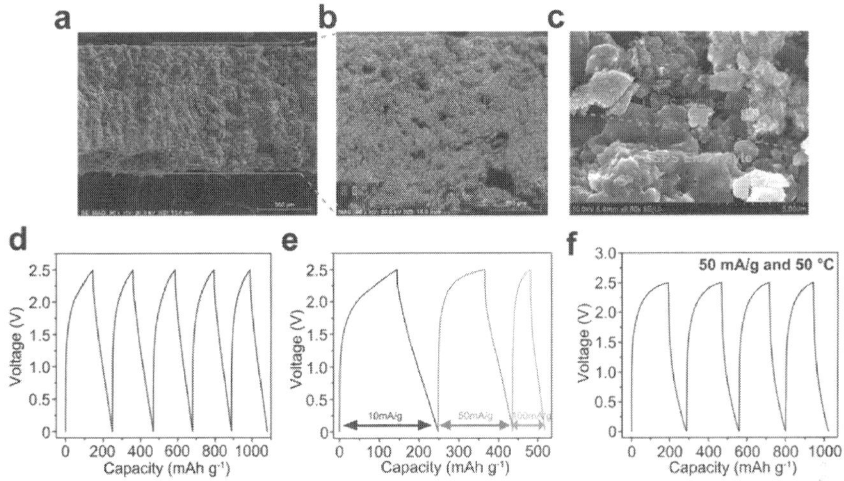

Figure 3.28. *Performances of all-solid-state LGPS-C/LGPS/LGPS-C batteries. (a) Cross section observed by scanning electron microscopy of an all-solid-state LGPS battery. (b) Elemental analysis in C (red) and sulfur (blue). (c) Observation of the LGPS-C electrode and LGPS electrolyte interface at a higher magnification. (d) Load–discharge curves of an LGPS battery in the range of 0.0–2.5 V from the second cycle at a current density of 10 mA/g. (e) Charge/discharge curves all-solid-state Li/LGPS/LGPS-C cells in the range of 0.0–2.5 V from the second cycle for different current densities and (f) cell charge–discharge curves all-solid-state Li/LGPS/LGPS-C cells in the range of 0.0–2.5 V from the second cycle at 50 °C under a current density of 50 mA/g. The mass ratio of the negative electrode to that of the positive electrode is set at 3 to balance the cell [HAN 15]. For a color version of this figure, see www.iste.co.uk/viallet/batteries.zip*

Tarhouchi *et al.* have shown that this concept can be transferred to the tin analog [TAR 15]. Since the low-potential electrochemical activity of the $Li_{10}SnP_2S_{12}$ phase has been confirmed by different characterizations, a cell where $Li_{10}SnP_2S_{12}$ would act as both solid electrolyte and negative electrode material was assembled using a composite positive $LiCoO_2/Li_{10}SnP_2S_{12}$ NEI1/VGCF electrode, commercial electrolyte $Li_{10}SnP_2S_{12}$ (NEI) and a composite $Li_{10}SnP_2S_{12}$ NEI1/VGCF negative electrode.

The cycling of the all-solid-state cell is shown in Figure 3.29. A disintercalation of 0.5 lithium from $LiCoO_2$, corresponding to the reaction of

1.3 lithium per mole of $Li_{10}SnP_2S_{12}$ at the negative electrode, occurred during the first charge. The voltage of the $Li_{0.5}CoO_2$ cell is 3.6 V, instead of 4.2 V versus Li^+/Li, meaning that the voltage at the negative electrode would be 0.6 V versus Li^+/Li. This potential could correspond to the phenomenon occurring at the second pseudo-plateau of 0.8 V versus Li^+/Li at 0.6 V versus Li^+/Li in Figure 3.29. After the first charge, only 0.05 lithium was reintercalated in $LiCoO_2$ during the first discharge, the capacity degrading during cycling.

No short circuits were observed during cycling, showing that the reaction at the negative electrode does not propagate through the electrolyte to the positive electrode.

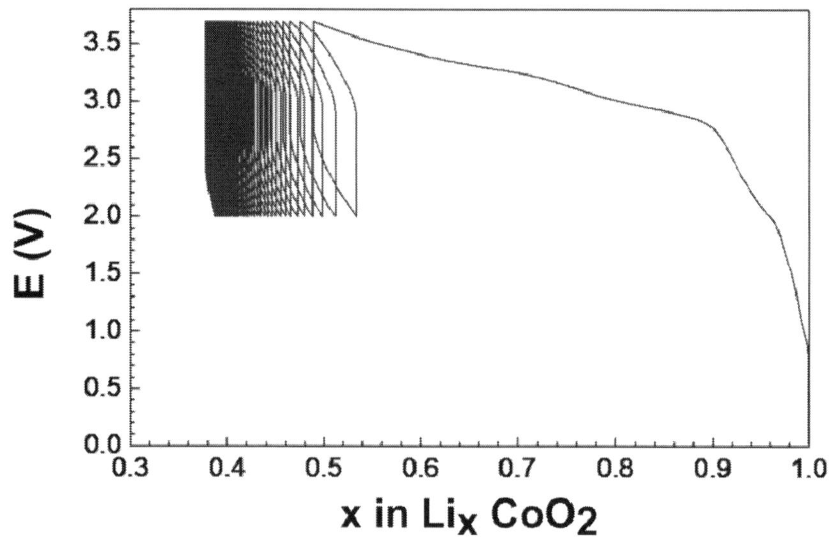

Figure 3.29. *Cycling curve of an all-solid-state $LiCoO_2$/ $Li_{10}SnP_2S_{12}$ NEI1 cell (64 µA/cm^2 – 1 Li in 20 hr) [TAR 15]*

It was verified that for a charged cell, the reaction proceeding to the negative electrode does not propagate through the electrolyte over time. An all-solid-state cell was charged and then placed in relaxation to observe the evolution of the cell voltage (Figure 3.30). For about 35 days, the cell voltage seemed to tend toward a constant value of 1.6 V. Then, the aging phenomenon appeared to be altered due to the significant decrease in voltage. This would be related to a propagation of the reaction occurring at the negative electrode throughout the cell.

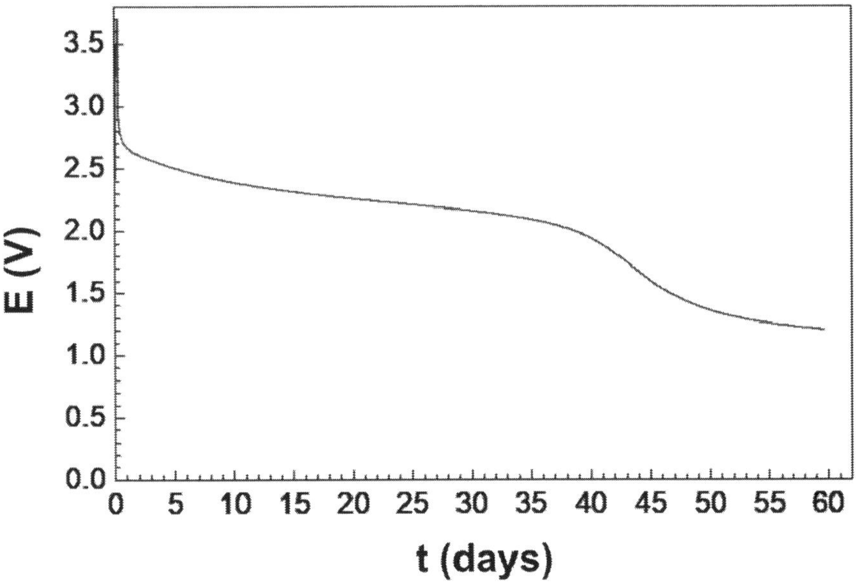

Figure 3.30. *Relaxation curve after the first charge of a $LiCoO_2/Li_{10}SnP_2S_{12}$ NEI1/VGCF cell [TAR 15]*

Very recently, in 2016, Kato *et al.* [KAT 16] showed remarkable battery performance with sulfur-based ion conductors $Li_{9.54}Si_{1.74}P_{1.44}S_{11.7}Cl_{0.3}$ and $Li_{9.6}P_3S_{12}$ crystallizing in the LGPS structure. $Li_{9.54}Si_{1.74}P_{1.44}S_{11.7}Cl_{0.3}$ has an exceptional ionic conductivity of 25 mS/cm at room temperature and $Li_{9.6}P_3S_{12}$ a remarkable potential stability close to 0 V versus Li metal.

The ionic conductivities as a function of the temperature of various electrolytes crystallizing in the LGPS structure as well as those of $Li_{9.54}Si_{1.74}P_{1.44}S_{1.7}Cl_{0.3}$ and $Li_{9,6}P_3S_{12}$ are grouped in Figure 3.31(a). The conductivity of $Li_{9.54}Si_{1.74}P_{1.44}S_{11.7}Cl_{0.3}$ (stars) is twice that of $Li_{10}GeP_2S_{12}$ (squares) and is the highest achieved to date in a solid electrolyte. This high conductivity can be explained by the 3D conduction paths (1D in the direction of the *c*-axis and and 2D in the plane *ab*) as indicated by the anisotropic thermal displacements of the lithium in Figure 3.31 (b) (spheres for Li(16h), Li(8f) and Li(4c)) and the electron density distribution (Figure 3.31(c)).

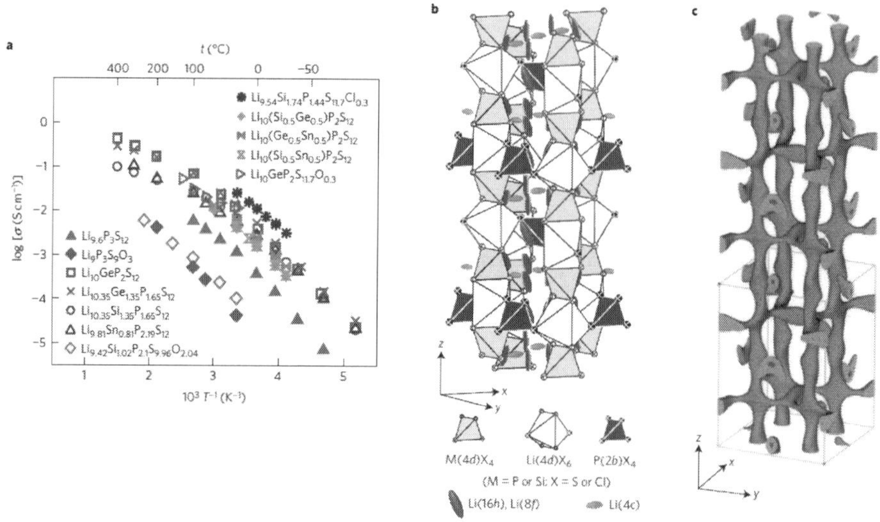

Figure 3.31. *Ionic conductivity and crystalline structure of $Li_{9.4}Si_{1.74}P_{1.44}S_{11.7}Cl_{0.3}$ [KAT 16]. For a color version of this figure, see www.iste.co.uk/viallet/batteries.zip*

The electrochemical stabilities with respect to the lithium metal of $Li_{9.54}Si_{1.74}P_{1.44}S_{11.7}Cl_{0.3}$, $Li_{9.6}P_3S_{12}$ and $Li_{10}GeP_2S_{12}$ were characterized from the charge–discharge measurements obtained on Li/solid electrolyte /$LiCoO_2$ cells. The Coulombic efficiency η of the different cells is calculated by taking the ratio of the discharge capacity to that of the load multiplied by 100. The charge and discharge capacities for the three systems are shown in Figure 3.32. The Coulombic efficiency of the cell assembled with $Li_{9.6}P_3S_{12}$, equal to 90%, is excellent. That of the cell assembled with $Li_{10}GeP_2S_{12}$, a material already identified as instable with respect to lithium, is only 61% and that with $Li_{9.54}Si_{1.74}P_{1.44}S_{11.7}Cl_{0.3}$ is only 39%. The decrease in the capacity indicates the presence of parasitic reactions during cycling, which causes the formation of a surface layer which blocks the transfer of lithium.

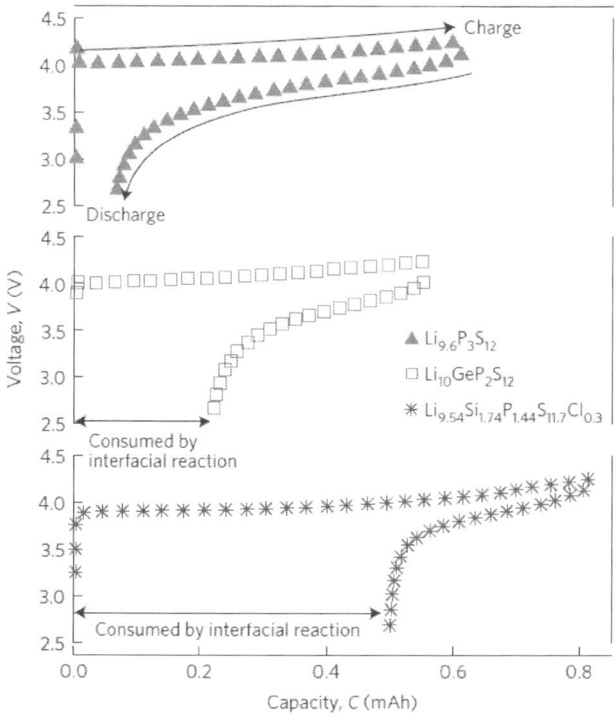

Figure 3.32. *Evolution of the voltage as a function of the charge and discharge capacity of the cells assembled with the solid electrolytes $Li_{9.54}Si_{1.74}P_{1.44}S_{11.7}Cl_{0.3}$, $Li_{9.6}P_3S_{12}$ and $Li_{10}GeP_2S_{12}$. Stability is evaluated by calculating the Coulombic efficiency η with η=(discharge capacity/charge capacity)×100 [KAT 16]. For a color version of this figure, see www.iste.co.uk/viallet/batteries.zip*

Three types of all-solid-state cells have been assembled to characterize the performance of cells with these solid electrolytes:

– [$Li_4Ti_5O_{12}$ + $Li_{10}GeP_2S_{12}$/$Li_{10}GeP_2S_{12}$/$LiNbO_3$-coated $LiCoO_2$ + $Li_{10}GeP_2S_{12}$] (normal cell);

– [graphite + $Li_{9.6}P_3S_{12}$/$Li_{9.6}P_3S_{12}$|$Li_{10}GeP_2S_{12}$/$LiNbO_3$-coated $LiCoO_2$ + $Li_{10}GeP_2S_{12}$] (high potential cell);

– [$Li_4Ti_5O_{12}$ + $Li_{9.54}Si_{1.74}P_{1.44}S_{11.7}Cl_{0.3}$/$Li_{9.54}Si_{1.74}P_{1.44}S_{11.7}Cl_{0.3}$/$LiNbO_3$-coated $LiCoO_2$ + $Li_{9.54}Si_{1.74}P_{1.44}S_{11.7}Cl_{0.3}$] (power cell).

In all cases, $LiCoO_2$ coated with a layer of $LiNbO_3$ was used as the positive electrode material ($LiNbO_3$-coated $LiCoO_2$).

The cycling performance of the all-solid-state cells is excellent and around 75% of the initial discharge capacity is retained after 500 cycles and at 100 °C (Figure 3.33) at which temperature conventional batteries do not operate.

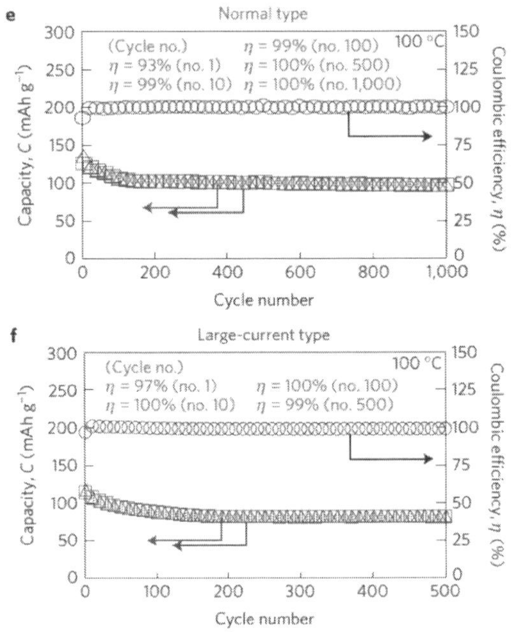

Figure 3.33. *Cycling performance of normal (top) or power (bottom) all-solid-state batteries [KAT 16]. For a color version of this figure, see www.iste.co.uk/viallet/batteries.zip*

The Ragone diagram, showing the performance of cells assembled by Kato *et al.* [KAT 16] as well as those of existing batteries and capacitors is presented in Figure 3.34. The Ragone diagram, commonly used to compare the performance of energy storage techniques (batteries, fuel cells, electrical storage devices, etc.) is a graph with logarithmic axes, representing, in the case of Figure 3.34, the specific power P in kW/kg as a function of the specific energy in Wh/kg. In general, the specific power is inversely proportional to the specific energy in electrochemical systems. In the study by Kato *et al.*, the power densities of $LiCoO_2$ coated with a layer of $LiNbO_3$ are higher than that of a conventional Li-ion battery, and even higher than those of materials used in supercapacitors.

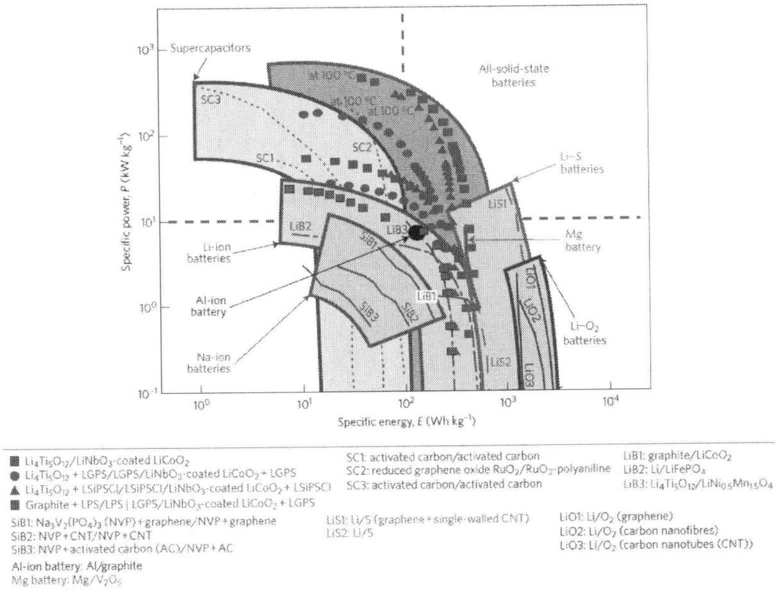

Figure 3.34. *Ragone diagram combining the performances of cells assembled by Kato et al. [KAT 16] and existing batteries and capacitors. The dashed lines indicate the specific energy $E = 10^2$ Wh/kg specific power $P = 10^2$ kW/kg. The all-solid-state cells prepared in this study simultaneously exhibit high powers and energies ($E = 10^2$ Wh/kg and $P = 10^2$ kW/kg, difficult to reach with conventional devices). For a color version of this figure, see www.iste.co.uk/viallet/batteries.zip*

3.1.6. *Understanding stability at the interfaces between the electrolyte and electrode materials*

The examples of all-solid-state batteries described earlier show how the phenomena at the interfaces are key to the performance of the devices using this technology. In addition to the studies carried out to limit reactivity, research was carried out to understand the reactions at the interfaces and the origin of this instability.

The use of lithium metal would allow the all-solid-state technology to be very competitive with that of Li-ion but no electrolyte has proven stable enough.

In 2015, Wenzel *et al.* demonstrated the formation of three different types of interface when a solid electrolyte is in contact with lithium metal [WEN 15]. In the first case, the one in which the interface is thermodynamically stable (Figure 3.35(a)), the solid electrolyte does not react at all with the metallic lithium and a sharp two-dimensional interface is formed. This case implies that the two phases are in thermodynamic equilibrium.

If the interface is thermodynamically unstable, the solid electrolyte reacts chemically with the metallic lithium and an interface is formed (Figure 3.35(b) and (c)).

If the reaction products between lithium and the electrolyte are electronic and ionic conductors, this new phase at the interface grows within the solid electrolyte, a layer known as the mixed conductive interface (MCI) (Figure 3.35(b)). This layer can allow the transport of electrons within the electrolyte (when only the ions need to be transported) and causes the battery to self-discharge.

If the products of the reaction between lithium and the electrolyte are not electronic conductors or if the electronic conduction is too weak, a new phase of limited thickness is formed at the interface, called the Solid Electrolyte Interface (SEI), and is comparable to the SEI observed in Li-ion batteries with liquid electrolytes.

The performances of an all-solid-state battery therefore strongly depend on the conduction properties of the layer formed at the interface.

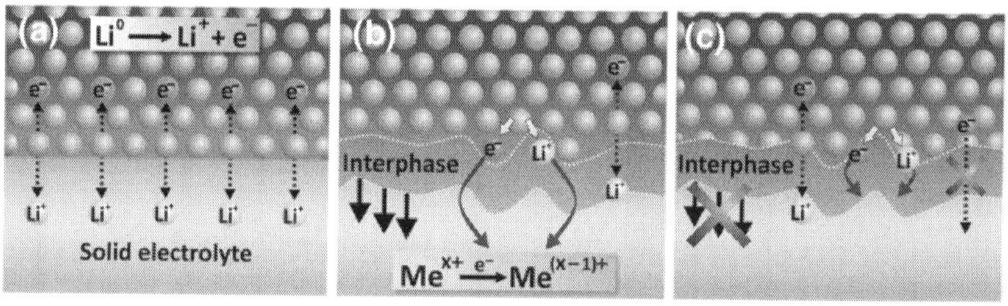

Figure 3.35. *Three types of interfaces existing between lithium metal and a solid electrolyte, conductor of Li^+ ions. a) Thermodynamically stable non-reactive interface, b) mixed conducting interphase (MCI), and c) reactive and metastable Solid Electrolyte Interface (SEI) [WEN 15]. For a color version of this figure, see www.iste.co.uk/viallet/batteries.zip*

Wenzel *et al.* developed a relatively simple technique in this same study in order to characterize this interface. This technique consists of depositing a thin layer of lithium on a surface of solid electrolyte and observing the reaction between the metal and the electrolyte just after the deposition of lithium by *in situ* X-ray photoelectron spectroscopy [WEN 15]. This technique made it possible to demonstrate the reduction of Ti^{4+} in Ti^{3+}, Ti^{2+} and even titanium metal at the lithium interface and the $Li_{0.35}La_{0.55}TiO_3$ electrolyte (LLTO).

Richards *et al.* developed a computational method to characterize the formation of these phases generated at the interface of the electrolyte and electrode materials in general, including metallic lithium [RIC 15].

The study deals with the reaction between the electrolyte and the positive electrode material when the cell is charged with the formation of a lithium-deficient decomposition layer within the electrolyte (Figure 3.36(a) as well as in the reaction between the electrolyte and lithium metal as a negative electrode (anode) (Figure 3.36(b)). The authors initially calculated the electrochemical stability of various binary lithium-containing materials (Figure 3.36(c)). This study shows that the halide anions are the most stable ones at high voltages, and the lithium nitrides and phosphides are the most stable at low voltages.

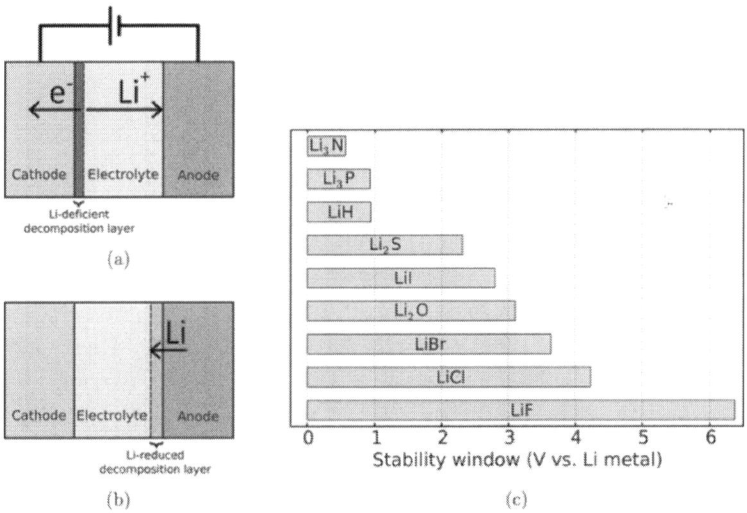

Figure 3.36. *Diagram of an all-solid-state cell showing (a) decomposition of the electrolyte at the electrolyte/positive electrode material interface during charging. (cathode) and (b) reduction of the electrolyte by lithium metal. (c) Areas of electrochemical stability of lithium binary compounds [RIC 15]. For a color version of this figure, see www.iste.co.uk/viallet/batteries.zip*

Richards *et al.* then calculated the electrochemical stability domains of a large number of materials used as solid electrolytes (Figure 3.37) and tabulated the expected decomposition products. This study confirms the instability of a large number of commonly used electrolytes.

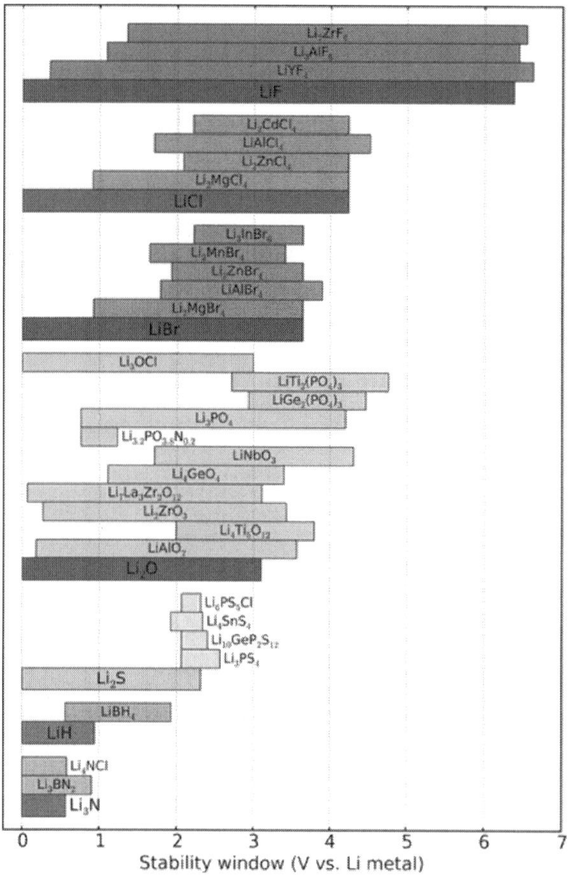

Figure 3.37. *Range of electrochemical stability of different solid electrolytes grouped by anion, with the corresponding binary [RIC 15]. For a color version of this figure, see www.iste.co.uk/viallet/batteries.zip*

3.1.7. *Summary*

The development of all-solid-state batteries has grown rapidly with research on sulfur-based electrolytes, in terms of developed electrolytes, but also on innovative cycling devices and characterization means (Figure 3.38).

The diversity of the batteries assembled in 2013 can be seen in Figure 3.38 grouping together the many negative and positive electrode materials used [TAT 13].

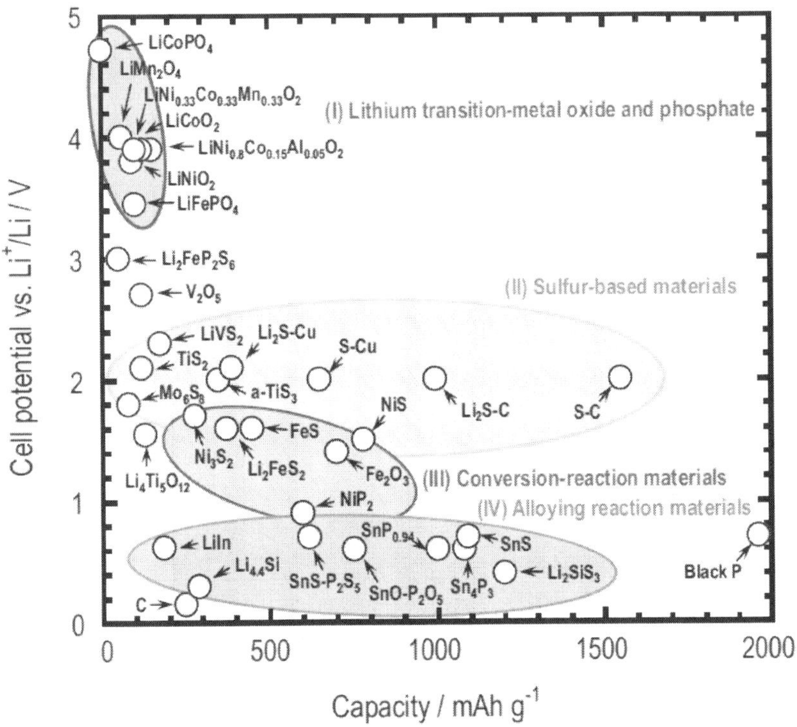

Figure 3.38. *Correlation between cell voltages (vs. Li+/Li) and reversible capacity (per gram of active material) in all-solid-state batteries using solid sulfur-based electrolytes [TAT 13]. For a color version of this figure, see www.iste.co.uk/viallet/batteries.zip*

3.2. Sodium monolithic "all-solid-state" batteries

The first all-solid-state sodium cells were assembled by cold pressing using the solid glass-ceramic Na_3PS_4 electrolyte by Hayashi *et al.* in 2012. The Na–Sn alloy was used as an anode and TiS_2 as the cathode. Figure 3.39 shows the charge–discharge curves of this all-solid-state cell at room temperature. The mean voltages of this cell were 1.6 V from the second to the 10th cycle, corresponding to the potential difference between the Na–Sn counter electrode (about 0.3 V relative to Na^+/Na) and the working electrode

TiS_2 (approximately 2.0 V with respect to Na^+/Na). The reversible capacity of the cell is approximately 90 mAh/g of TiS_2 and over 10 cycles. The Osaka team showed for the first time the operation, at ambient temperature, of an all-solid-state rechargeable battery using sodium.

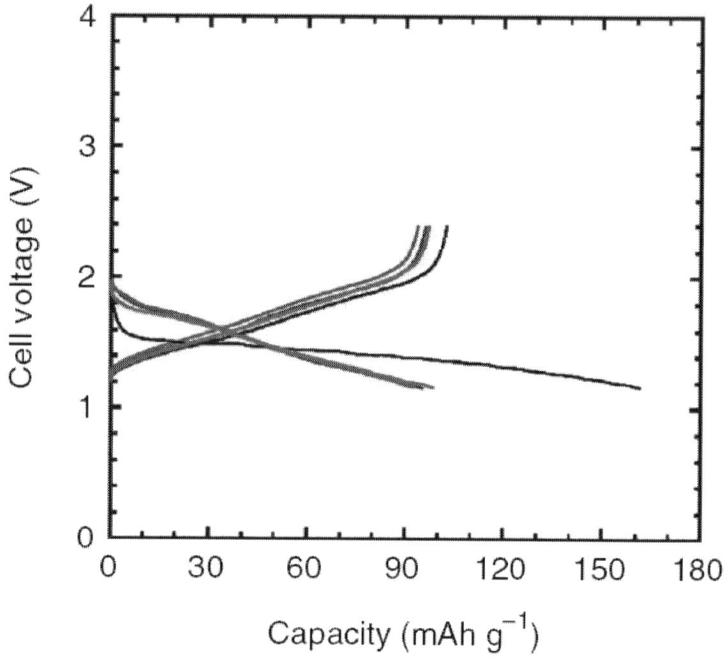

Figure 3.39. *Charge–discharge curves at 25 °C under a current density of 0.013 mA/cm^2 of a sodium all-solid-state rechargeable Na–Sn/Na_3PS_4–glass-ceramic/TiS_2 cell. The black, red, blue and green lines represent the first, second, third and 10th cycle, respectively. For a color version of this figure, see www.iste.co.uk/viallet/batteries.zip*

In 2014, after optimizing the preparation of Na_3PS_4 and doubling its conductivity with respect to the initial syntheses, Hayashi *et al.* assembled an all-solid-state $Na_{15}Sn_4/Na_3PS_4$ glass-ceramic/composite-$NaCrO_2$ [HAY 14]. The positive electrode is a composite obtained by mixing $NaCrO_2$, glass-ceramic Na_3PS_4 and acetylene black (AB) with a mass ratio of 4:6:1. The counter electrode is also a composite, $Na_{15}Sn_4$–AB, obtained by the mechanical milling of Na, Sn and AB; the weight ratio of $Na_{15}Sn_4$:AB being 20:3. The powders of the working electrode and the solid electrolyte were placed in a 10-mm-diameter polycarbonate tube and pressed together under a pressure of 70 MPa, composite counter-electrode $Na_{15}Sn_4$–AB was then placed on the surface of the solid electrolyte side of the bilayer pellet and a pressure of 360 MPa was applied to all three layers. The electrochemical characterizations were carried out at constant current densities of 0.013, 0.064 or 0.13 Ma/cm^{-2} in the voltage range of 1.2–4.0 V, at room temperature. After a first charge capacity of 130 mAh/g, the all-solid-state Na-Sn/$NaCrO_2$ battery shows a capacity close to 60 mAh/g of $NaCrO_2$ over 15 cycles (Figure 3.40) [HAY 14].

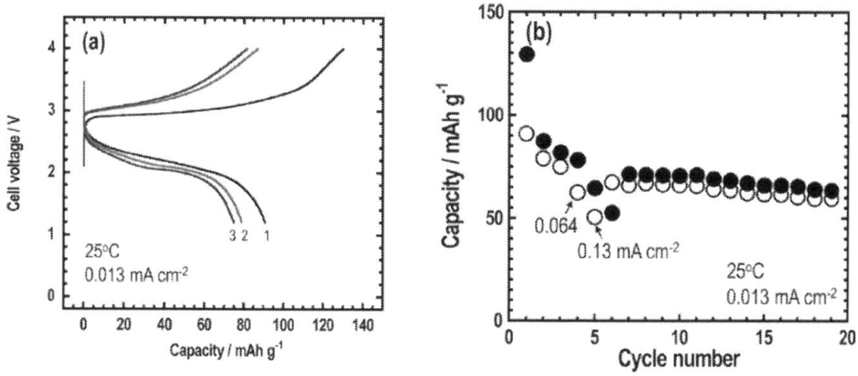

a) b)

Figure 3.40. *(a) The three first charge–discharge curves and (b) cycling performance of an all-solid $Na_{15}Sn_4$/ glass-ceramic cubic-Na_3PS_4 /$NaCrO_2$ battery. The full and open circles in the figure represent charge and discharge capacities, respectively [HAY 14]. For a color version of this figure, see www.iste.co.uk/viallet/batteries.zip*

In 2015, Hayashi *et al.* published the assembly of an all-solid-state battery based on amorphous *a*-TiS_3 [TAN 15]. These $Na_{15}Sn_4$/Na_3PS_4 glass-ceramic/*a*-TiS_3 cells deliver a high capacity of more than 300 mAh/g of TiS_3 in the first charge–discharge cycle. Capacity gradually decreased to 100 mAh/g at the 10th cycle (Figure 3.41(a)). The Nyquist diagrams of the cell showed that the resistance of the composite *a*-TiS_3 electrode increases as the charge and discharge cycles continue. The X-ray diffraction diagrams of the electrode before and after the first cycle showed that the *a*-TiS_3 compound remains amorphous during the charge–discharge process. The addition of acetylene black to the composite as a conductive additive allowed the cells to retain a high capacity of more than 300 mAh/g for five cycles (Figure 3.41(b)).

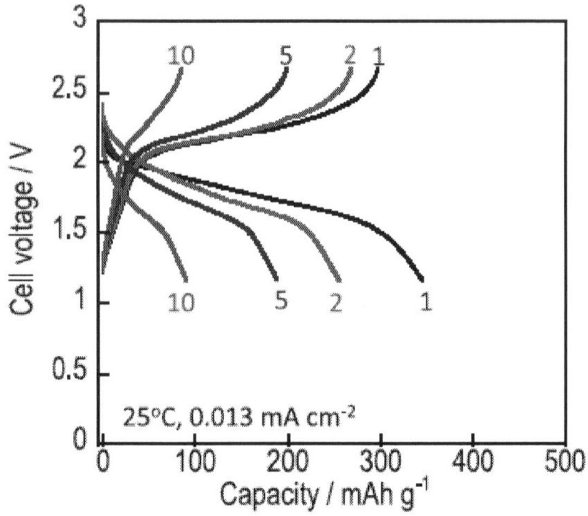

a)

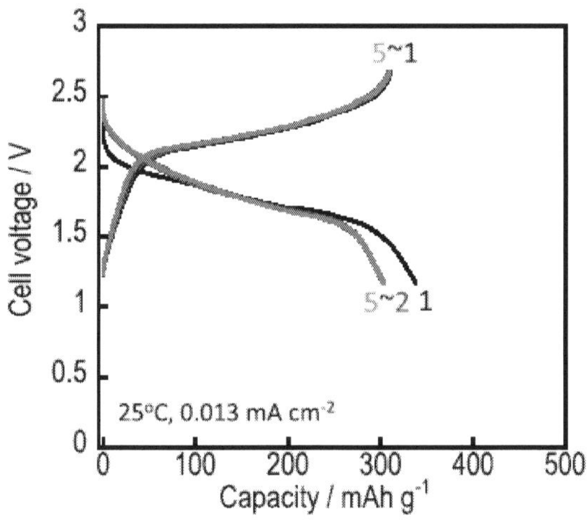

b)

Figure 3.41. *Charge–discharge curve of an all-solid-state $Na_{15}Sn_4/Na_3PS_4$ glass-ceramic/a-TiS_3 cell without the addition (a) and with the addition of acetylene black (b). For a color version of this figure, see www.iste.co.uk/viallet/batteries.zip*

Symmetrical cells using $Na_3V_2(PO_4)_3$ (denoted NVP) at the negative electrode and the positive electrode, composite NVP/Na_3PS_4/composite NVP, were successfully assembled by Boulineau *et al.* [BOU 13]. $Na_3V_2(PO_4)_3$ can effectively be used as both negative and positive electrodes [YUB 15, LAL 15]. The mechanisms of reactivity of this compound are the following:

$$NaV_2^{IV}(PO_4)_3 \Leftrightarrow Na_3V_2^{III}(PO_4)_3 \Leftrightarrow Na_4V_2^{II,III}(PO_4)_3$$

The oxidation of V^{3+} into V^{4+} takes place at a voltage of 3.4 V versus Na^+/Na with a theoretical capacity of 117.6 mAh/g, and the reduction of V^{3+} into V^{2+} takes place at 1.6 V versus Na^+/Na with a capacity of 58.8 mAh/g. The cycling curves of a liquid electrolyte cell of this compound as a negative and positive electrode are shown in Figure 3.42(a). The amphoteric nature of this compound allows for the assembly of symmetrical NVP/NVP cells delivering an average voltage of 1.8 V, i.e. a theoretical specific energy of 211.68 Wh/kg [CAR 83, KNU 83]. The galvanostatic cycling curve of such a symmetrical cell in liquid electrolyte is presented in Figure 3.42(b) [LAL 15].

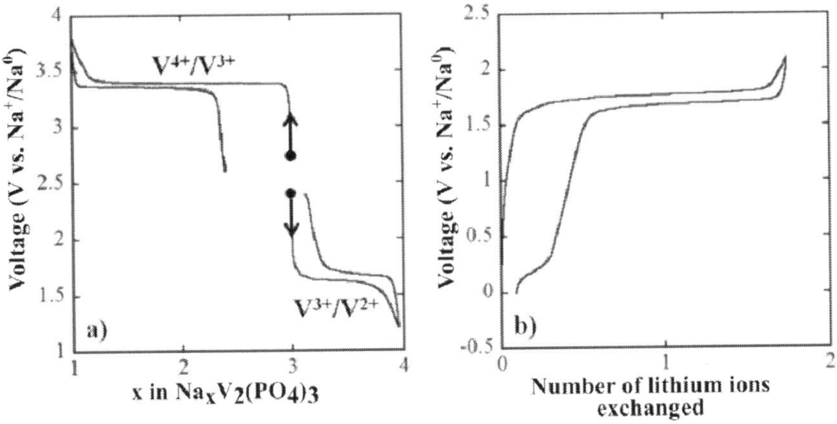

Figure 3.42. *a) Galvanostatic cycling curve of NVP as positive and negative electrode materials. b) Galvanostatic cycling curve of a symmetrical NVP/NVP cell. Batteries cycled at 25 °C, at C/20, in Swagelok®, electrolyte: 1 M $NaClO_4$ in PC. For a color version of this figure, see www.iste.co.uk/viallet/batteries.zip*

A symmetrical cell with Na_3PS_4 as the solid electrolyte has therefore been studied. The composition used for each of the electrodes is 27% NVP/63% Na_3PS_4/10% VGCF. The battery was assembled by depositing 10 mg of each electrode on each side of 80 mg of electrolyte Na_3PS_4, the assembly is shaped at 375 MPa under a uniaxial press in a 10-mm-diameter resin matrix.

The theoretical capacity of this cell is 58.8 mAh/g, i.e. one sodium exchanged at 1.8 V versus Na^+/Na. The cycling results, shown in Figure 3.43, show that 0.9 sodium is exchanged during charging i.e. 90% of the theoretical capacity [BOU 13]. However, due to a high polarization of 500 mV, only 35 mAh/g is obtained during the discharge, i.e. 60% of the theoretical capacity. The capacity drops rapidly to 10 mAh/g due to an increase in the polarization of the cell during cycling and increases again after 20 cycles. Despite this instability of the discharge capacity, the battery has a Coulombic efficiency of 98% over more than 30 cycles.

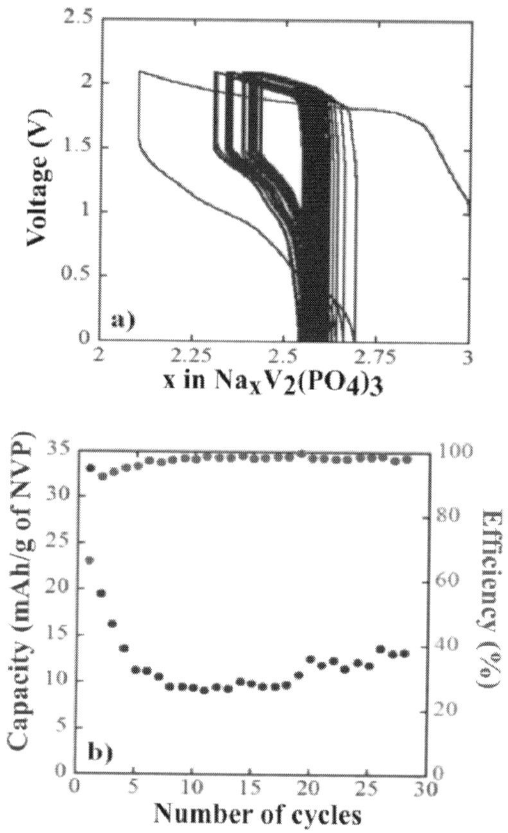

Figure 3.43. *a) Galvanostatic cycling curve of a symmetrical NVP/Na_3PS_4/NVP cell, at C/25, at room emperature. The composition of the two electrodes is 28.5%NVP/66.5%Na_3PS_4/10%VGCF. b) Associated capacity retention curve. For a color version of this figure, see www.iste.co.uk/viallet/batteries.zip*

This symmetrical sodium "all-solid-state" sodium battery assembled using the amphoteric $Na_3V_2(PO_4)_3$ material delivers a capacity of 15 mAh/g after 30 cycles, i.e. 26% of the theoretical capacity. Nevertheless, these are only preliminary tests and optimization of electrode composition would be necessary to improve performance.

3.3. "All-solid-state" Li–S batteries

As mentioned earlier, the use of a solid electrolyte makes it possible to obtain better performances with certain materials in comparison with conventional batteries. Thus, apart from the safety aspect, "all-solid-state" batteries appear to be a solution to the limitations imposed by liquid electrolytes. Much research has been done on the Li–S system in order to highlight the value of solid electrolytes, a system that still poses many problems today.

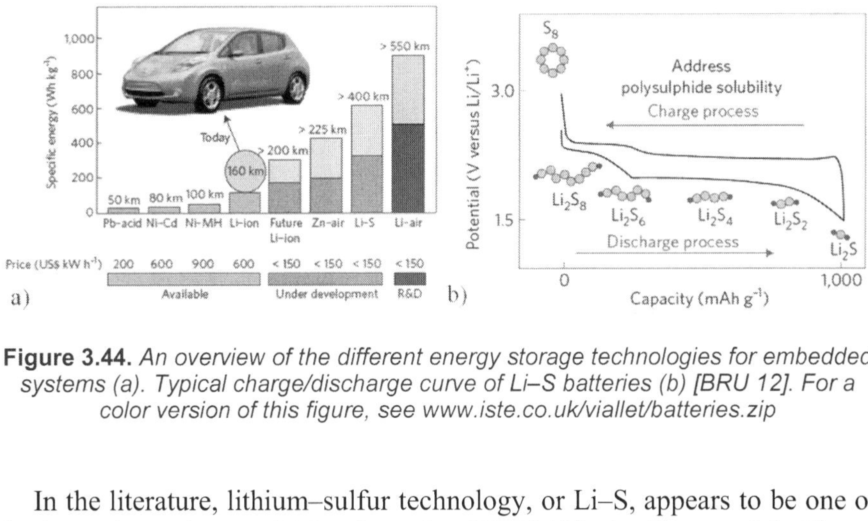

Figure 3.44. *An overview of the different energy storage technologies for embedded systems (a). Typical charge/discharge curve of Li–S batteries (b) [BRU 12]. For a color version of this figure, see www.iste.co.uk/viallet/batteries.zip*

In the literature, lithium–sulfur technology, or Li–S, appears to be one of the best alternatives to Li-ion batteries [BRU 12]. As illustrated in Figure 3.44(a), electric vehicles equipped with Li–S batteries would have a range of 400 km, almost twice the current range of electric vehicles using Li-ion technology. In addition, the cost of these batteries would be four times lower, under 150$ instead of 600$ per kW/h. Unfortunately, dissolution of the active substance during cycling is an issue, which does not currently allow this technology to meet the expectations of manufacturers. Indeed, as seen in Figure 3.44(b), the reduction of sulfur S to Li_2S requires the formation of electrolyte-soluble polysulfide chains Li_2S_x ($1 \leq x \leq 8$), which react with the lithium counter-electrode; this mechanism called "redox shuttle" results in a deterioration in electrochemical performance. Li–S "all-solid-state" batteries are designed to limit this dissolution phenomenon and thus improve the properties of these batteries.

The use of elemental sulfur S_8 for an "all-solid-state" battery application poses many problems because of the extremely insulating nature of this compound ($\sigma = 5 \times 10^{-30}$ S/cm). While most composite electrodes that we have seen previously are obtained by simple mixing with mortar, the preparation of sulfur-based composite electrodes must be optimized. In the literature, two techniques are primarily used: mechanical milling and gas phase mixing.

The first, proposed in 2004 by Machida [MAC 04] and optimized by Hayashi *et al.*, was that of mechanical milling. To compensate for the low conductivity of sulfur, a S/CuS mixture obtained by mechanical milling of sulfur and copper was used in the first batteries. The latter, mixed in a mortar with the $80Li_2S/20P_2S_5$ electrolyte and acetylene black, has an excellent capacity retention of 650 mAh/g for 20 cycles, at a current density of 64 μA/cm², results from several reaction mechanisms involving CuS, S and Li_xCuS alloy [HAY 03c]. The Li_2S/Cu mixture has also been studied but has a lower electrochemical performance compared to its counterpart using elemental sulfur [HAY 08b].

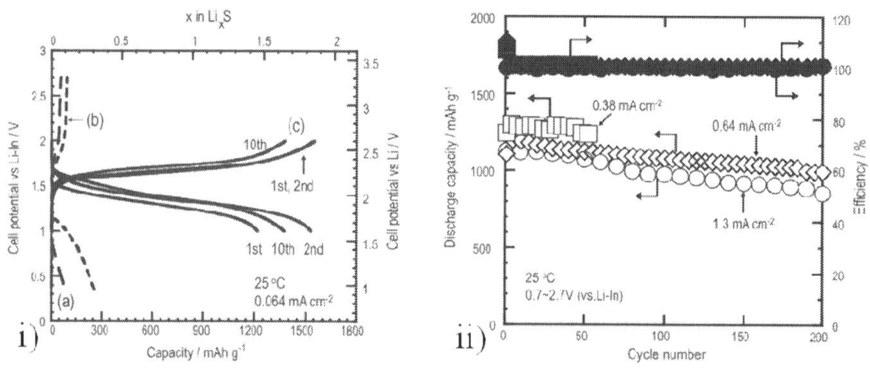

Figure 3.45. *(i) Influence of the preparation of the composite on the capacity delivered via the mortar (a), the mortar + planetary mill (b) and the planetary mill (c). (ii) Capacity retention of the composite prepared by mechanical grinding*

The use of copper was quickly abandoned because it was too heavy and too expensive, in favor of a more conventional mixture comprising the active substance, the electrolyte and a conductive additive. In 2011, Nagao *et al.* presented an "all-solid-state" (Li–In)/S battery with much higher electrochemical properties than batteries using an organic liquid electrolyte [NAG 11b]. The latter, based on the simultaneous mixing of sulfur, $80Li_2S$–$20P_2S_5$ electrolyte and acetylene black by mechanical milling (25:50:25, respectively), delivers a capacity of 1450 mAh/g after ten cycles at a current density of 64 $\mu A/cm^2$, as shown in Figure 3.45(i).

Nevertheless, during charging, the oxidation of Li_2S obtained by the reduction of sulfur is accompanied by the oxidation of Li_2S contained in the electrolyte. This phenomenon, observable on the cycling curve, where the capacity of the first charge is substantially greater than that of the first discharge, leads to a rapid decrease in the capacity by deterioration of the composite electrode. As seen in Figure 3.45(ii), the increase in the cycling rate has made it possible to limit the oxidation of the electrolyte and has led to a capacity retention of 1000 mAh/g for 200 cycles at a current density of 1.3 mA/cm^2. It is also interesting to note that the shape of the cycling curve in Figure 3.45(i) is substantially different from that of Figure 3.44(b), suggesting different reaction mechanisms when using a solid electrolyte in which the polysulfides are not soluble.

In 2012, Nagao *et al.* presented the first Li–S "all-solid-state" battery, using a lithium deposit, made by PVD to stabilize the interface between lithium and electrolyte [NAG 11]. This battery delivers a capacity of 1000 mAh/g for 20 cycles at a current density of 64 $\mu A/cm^2$ at C/15 (1 Li^+ in 15 hr). A second technique for preparing the composite electrode was presented by Kanno *et al.* and consists of a solid–gas mixture between the sulfur and a porous carbon. The aim of this technique is to mix the sulfur and the carbon by mechanical grinding and then heat-treat them in a tube sealed under vacuum in order to evaporate the sulfur and to improve the homogeneity of the mixture [KOB 08, NAG 13]. As illustrated in Figure 3.46(a) and (b), the composite electrode prepared in this manner exhibited a capacity retention of 400 mAh/g for five cycles at a current density of 0.13 mA/cm^2, compared with 40 mAh/g delivered by the composite prepared by simple planetary grinding.

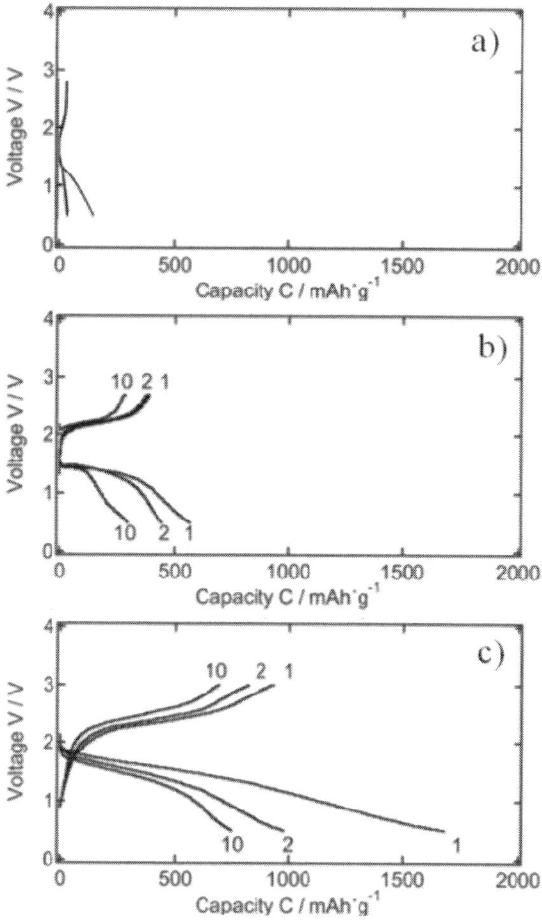

Figure 3.46. *The charge/discharge curve of composite electrodes prepared by mechanical grinding (a) and by gas–solid mixture: sulfur–acetylene black (b) and sulfur-CMK 3 (c) [NAG 11]*

In order to improve the electrochemical performance of these batteries, acetylene black has been replaced by CMK-3, a carbon with a two-dimensional cylindrical rod structure. The structural organization of this carbon allows more intimate contacts between the electrolyte and the sulfur-CMK-3 composite, and therefore, better electrochemical performances (Figure 3.46(c)).

In 2015, Chen and Adams reported the performance of an all-solid-state S/Li_6PS_5Br/In–Li battery. The positive electrode is a composite, a mixture of S, super carbon P and Argyrodite, obtained in two steps. Sulfur and carbon are first mixed with the planetary mill at 500 rpm for 5 hr, the initially crystallized sulfur becomes amorphous and intimately mixed with carbon with particle sizes of less than 100 nm. The Argyrodite Li_6PS_5Br is then added to the S–C mixture and a second grinding at 500 rpm for 30 min is carried out. 100 mg of solid electrolyte and 7.5 to 15 mg of composite electrode (S mass fixed to 3 mg) are placed in a 13 mm diameter matrix and pressed under 6000 kg/cm^2. The composite electrode then has a thickness of 30–40 μm and the electrolyte has a thickness of 300 μm. An In–Li sheet is then placed on the electrolyte side, the full-cell is then placed in a Swagelok device.

Electrochemical performances were characterized at C/10, i.e. a current density of 167 mA/g of sulfur (0.38 mA/cm^2), between 0.4 and 2.4 V versus In–Li (1.0–3 V versus Li^+/Li). A battery with 20% mass of S has an initial discharge capacity of 1355 mAh/g (i.e. 3.06 mAh/cm^2) (Figure 3.47).

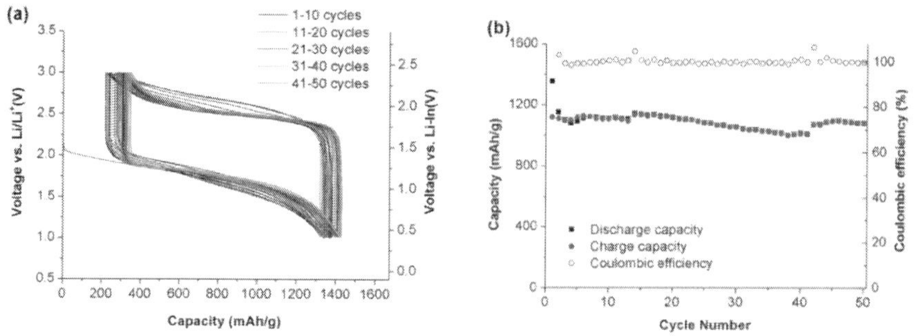

Figure 3.47. *a) Charge–discharge curves and (b) charge and discharge capacities and Coulombic efficiency of an all-solid-state S/Li_6PS_5Br/In–Li batter over 50 cycles at a current density of 167 mA/g of active S (corresponding to C/10) between 1.0 and 3.0 V versus Li^+/Li. (Mass ratio of the components of the composite: S:Li_6PS_5Br:C = 20:70:10.) For a color version of this figure, see www.iste.co.uk/viallet/batteries.zip*

The large drop in capacity on the second cycle to a value of 1150 mAh/g is probably related to the large volume variation of the active material during cycling (S+2Li↔Li_2S). From the second to the 50th cycle, the capacity loss is 0.13% per cycle, the Coulombic efficiency is therefore high (94%), the battery delivering a reversible capacity of 1080 mAh/g for 50 cycles at C/10 (Figure 3.48).

Increasing the sulfur content in the composite results in an increase in the initial discharge capacity, but the reversible discharge capacity is significantly higher for lower sulfur contents, thus confirming the key role changes in volume of the active material have on the loss of capacity (Figure 3.48).

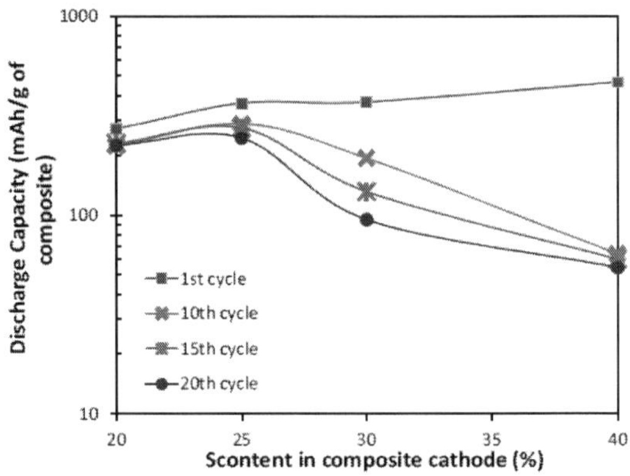

Figure 3.48. *Discharge capacity of all-solid-state batteries with cathodes having different sulfur contents (20, 25, 30 and 40%). The capacity is given in mAh/g of composite in order to better measure the specific energy of the cells. For a color version of this figure, see www.iste.co.uk/viallet/batteries.zip*

4

Monolithic "All-Solid-State" Batteries Using Solid Oxide Electrolytes

As we have just seen, sulfur-based glasses and glass-ceramics are the materials of choice for making high-performance "all-solid-state" batteries due to their ease of fabrication.

Despite the large number of ceramic electrolytes, such as the NASICONs or garnets, very few batteries using these compounds are reported. This limited number of studies can mainly be explained by the difficulty of preparing favorable interfaces for good cyclability in these cells. Nevertheless, several groups have succeeded in developing new techniques for fabricating monolithic solid-state ceramic batteries.

4.1. Silver "all-solid-state" battery technology

In 1998, Venkateswarlu and Satyanarayana developed and tested a silver-based solid-state primary battery [VEN 98]. The electrolyte was a glass with composition $AgI–Ag_2O–(SeO_2+As_2O_5)$, called SSA, which has a conductivity of 2.63×10^{-2} S /cm at 30 °C and an activation energy of 0.22 eV [OHA 10]. These solid electrolytes exhibit ionic conductivity and high stability under varying temperatures.

Two types of cell were assembled and characterized:

Anode (Ag_SE): solid electrolyte (SSA):cathode (I:C) cell (1)

Anode (Ag_SE): solid electrolyte (SSA):cathode (I_C:SE) cell (2) with Ag being the silver powder, SE the solid electrolyte, I the iodine, and C the graphite powder.

The open circuit voltages (or OCV, the voltage that designates the equilibrium voltage of the battery at rest), measured immediately after the cells were assembled, were 686 and 683 mV for cells (1) and (2), respectively (Figure 4.1). The discharge capacities were reported as 1.37 and 2.12 mAh, and the specific energies as 0.68 and 1.06 W/kg, for cells (1) and (2), respectively. These results confirm that the addition of electrolyte in the I+C cathode improves the interfacial properties and hence the battery performance.

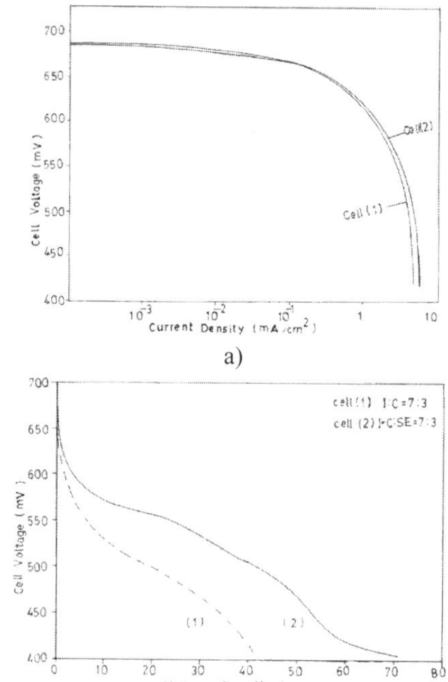

Figure 4.1. *(a) Polarization and (b) discharge curves for cells (1) and (2). Anode (Ag_SE): solid electrolyte (SSA): cathode (I:C) cell (I). Anode (Ag_SE): solid electrolyte (SSA): cathode (I_C:SE) cell (2)*

In 2003, Das *et al.* studied the operation of solid-state batteries with silver phosphate glass electrolytes doped with different cadmium and zinc halides [DAS 03]. Metallic silver was used for the negative electrode and graphite mixed with iodine was used for the positive electrode. Electrochemical characterization demonstrated a high ion transport number, close to 98.5%, for the glasses doped with cadmium chloride, which had a conductivity of around 4.64×10^{-5} S/cm.

In 2004, silver-based glass electrolytes with formulae x [0.75AgI: 0.25AgCl]:(1 x) [Ag_2O: WO_3] were prepared by Agrawal *et al.* from precursor salts using the melt-quenching technique [AGR 04]. A conductivity of 4×10^{-3} S/cm was obtained at ambient temperature for the stoichiometry 0.7[0.75AgI:0.25AgCl]:0.3[Ag_2O:WO_3]. These electrolytes were used to assemble all-solid-state batteries with metallic silver at the negative electrode and a mixture of carbon and iodine at the positive electrode.

In 2012, Suthanthiraraj and Sarumathi studied silver phosphate ($Ag_4P_2O_7$) based solid electrolytes and variable concentrations of antimony iodide (SbI_3), prepared by melt-quenching [SUT 12]. A maximum ionic conductivity of 4.1×10^{-4} S/cm was obtained for a 60% molar concentration of SbI_3. Above this value, the conductivity decreases as a result of the formation of cross-links in the pyrophosphate network, which hinders the migration of silver ions. Nanoscale particles of silver iodide (AgI), observed in the electrolyte under FESEM analysis, were thought to cause the improvement of the ionic conductivity. These electrolytes have been used to assemble solid-state batteries with negative and positive electrodes composed of silver iodide mixed with carbon fibers and phenothiazine, respectively.

Masoud *et al.* studied the potential application of silver-based borate glasses in solid-state battery systems [MAS 13]. Borate glasses, especially when doped with silver iodide (AgI), are claimed to exhibit high ionic conductivity. The borates used in these experiments had formulae xAgI–(1–x) [$_{0.67}Ag_2O$–$_{0.33}B_2O_3$], with x equal to 0.4, 0.5, 0.6, 0.7 and 0.8. The positive electrode was a mixture of carbon and iodide in a 30:70 ratio by mass, while the negative electrode was made up of glass and silver powder in a 67:33 ratio by mass. The charge–discharge characteristics showed optimal performance for the composition x = 0.6. For this composition, the borate glass exhibits a conductivity of 6.73×10^{-4} S/cm at ambient temperature.

Delaizir *et al.* developed a silver-based solid-state battery assembled in one step using the spark plasma sintering (SPS) technique [DEL 12a]. At ambient temperature, silver- and copper-based ionic conductors tend to exhibit high ionic conductivities: 0.05 S/cm for glassy $Ag_6I_4WO_4$ and 0.34 S/cm for crystalline $Rb_4Cu_{16}I_7Cu_{14}$. The composite electrodes used were obtained by mechanical milling $Ag_{0.7}V_2O_5$ and $Ag_6I_4WO_4$ (1:1 mass ratio) for around 2 hr. The performance of a symmetric solid-state composite SE-$Ag_{0.7}V_2O_5$/$Ag_6I_4WO_4$ (SE)/composite SE-$Ag_{0.7}V_2O_5$ cell, stable in open air, was compared to that of cold-pressed solid-state batteries with thick composite electrodes (thicknesses greater than 400 µm). These cold-pressed batteries exhibit limited electrochemical properties due to significant polarization in the cell associated with poorly defined electrode/electrolyte interfaces (Figure 4.2(a)). On the other hand, thick batteries obtained by SPS exhibit excellent reversibility without requiring pressure to be applied throughout cycling (Figure 4.2(b)). This improvement can be explained by well-defined interfaces and good mechanical stability preserved during cycling. The electrochemical formation of $Ag_xV_2O_5$ during charge–discharge cycles does not lead to drastic volume change. The electrode/electrolyte interfaces are preserved.

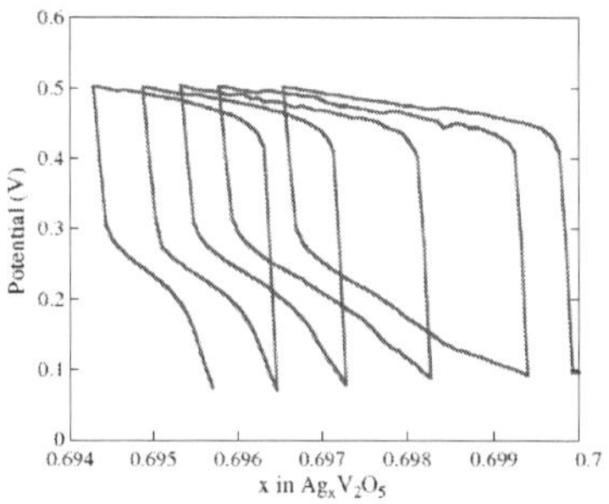

a)

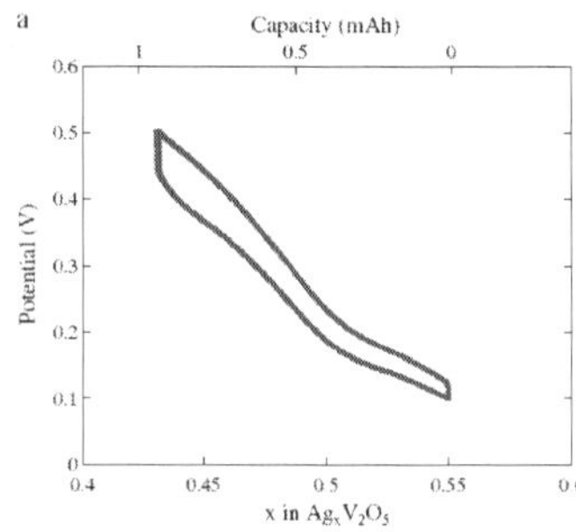

b)

Figure 4.2. *Charge–discharge curves at 20 °C for a current of 50 µA for batteries assembled by cold pressing (a) and by spark plasma sintering (b) at 20 °C*

4.2. Li-ion "solid-state" battery technology

A major advance in the assembly of these batteries was brought about by the use of spark plasma sintering (SPS) as a fabrication technique. This technique allowed monolithic batteries with thicknesses between 0.9 and 1.6 mm to be achieved, based upon an "all-NASICON" structure and exhibiting comparable electrochemical performance to conventional Li-ion batteries using liquid electrolytes [ABO 11, DEL 12b].

The electrolyte $Li_{1.5}Al_{0.5}Ge_{1.5}(PO_4)_3$ (written as LAG or LAGP in different publications) was chosen for the solid electrolyte due to its relatively high conductivity of 3×10^{-4} S/cm at 25 °C, and its attractive window of electrochemical stability (between 1.8 and 7 V *vs.* Li^+/Li).

$LiFePO_4$, denoted as LFP, which exhibits a reversible capacity of 170 mAh/g at a potential of 3.5 V versus Li^+/Li, was found to be compatible with LAG during SPS assembly, as well as $Li_3V_2(PO_4)_3$ (denoted as LVP). LVP can be used as the positive electrode material (with a reversible capacity of 120 mAh/g between 3 and 4.2 V vs. Li^+/Li and 140 mAh/g between 3 V and 4.65 V vs. Li^+/Li) or the negative electrode material (80 mAh/g, corresponding to the insertion/deinsertion of 1.2 Li^+ per mole of LVP, between 1.8 and 3 V vs. Li^+/Li). These materials were chosen for their thermal stability, which is needed to withstand the fabrication at temperatures above 650 °C, and their window of electrochemical activity, with LAGP reducing below 1.8 V versus Li^+/Li. Super P carbon (Timcal, labeled as Csp) was used as a conducting additive.

The formulation of the composite electrodes was optimized at a value of 25/60/15 (active material/LAG/carbon molar ratio) in order to guarantee ionic and electronic percolation [DEL 12] (Figure 4.3(a)). A carbon content of more than 10% by mass is needed to obtain an electronic conductivity of 1 S/cm, and a study of the influence of the LVP/LAG ratio (1/2, 1/1 and 2/1) eventually led to the 25/60/15 formulation.

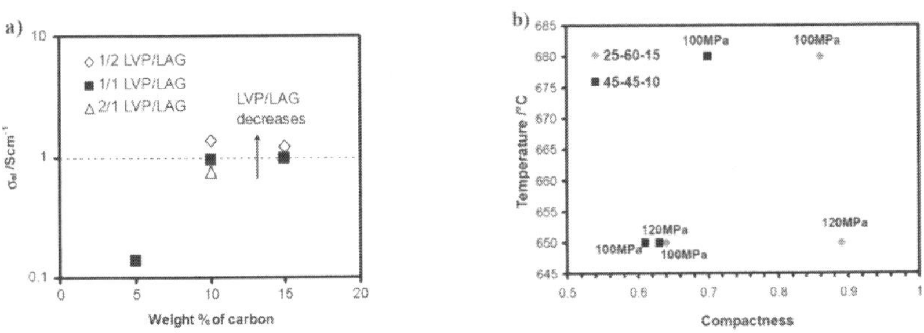

Figure 4.3. *a) Electronic conductivity of the LVP/LAGP/Csp composite electrodes. b) Influence of the temperature and pressure (2 min exposure) on the composite electrode compactness for different formulations [DEL 12]*

The pressure (75 MPa or 100 MPa) and temperature (500 to 850 °C) conditions lead to very different values of compactness, ranging between 57 and 93% (Figure 4.4(a)). The ionic conductivity of LAG alone in a cell was determined from Nyquist plots (Figure 4.4(b)). The first circle at high frequencies corresponds to the grain conductivity and the second to the grain boundary conductivity. The conductivity follows the Arrhenius equation and the activation energy calculated from the slopes of the curves equals 0.39 eV, a value independent of the compactness, in contrast to the conductivity (Figure 4.4(c)). The graph of isothermal conductivity as a function of the compactness (Figure 4.4(d)) allowed for the conductivity for a pellet with 100% compactness to be extrapolated as 2.8×10^{-4} S/cm, a value in good agreement with the literature.

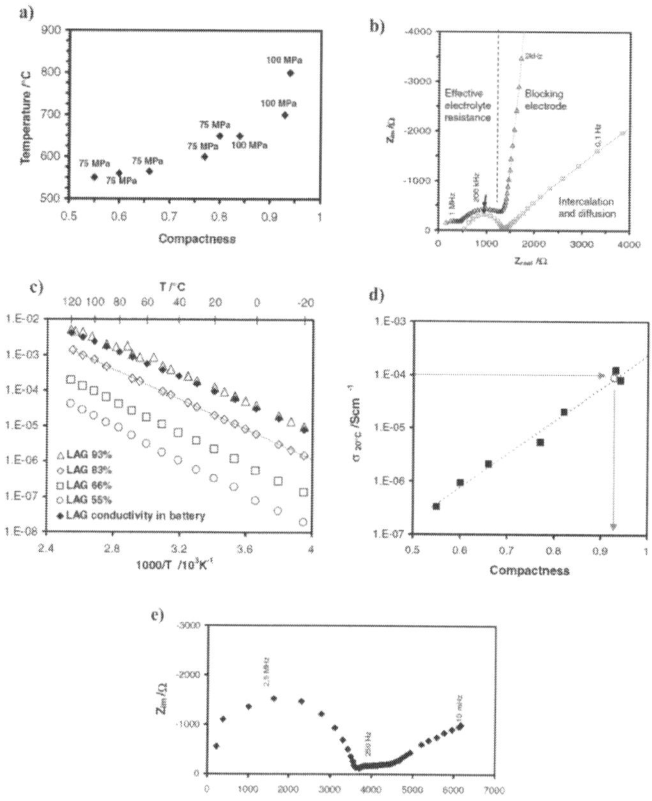

Figure 4.4. *a) Influence of the pressure and temperature of the SPS treatment on the compactness of the LAG electrolyte. b) Impedance spectra at ambient temperature for an LVP/LAG/LVP cell (squares) and for LAG (triangles) densified by SPS. c) Ionic conductivity in LAG as a function of temperature for different compactnesses (open symbols) and in an LVP/LAG/LVP cell (full symbols). d) Isothermal variation in conductivity as a function of compactness. This trace was used as a reference for determining the conductivity simply from the compactness (open circle). e) Impedance spectrum of a gold-coated LVP pellet with a compactness of 76% at 130 °C [DEL 12]*

LFP/LAG/LVP and symmetrical LVP/LAG/LVP batteries could be assembled by SPS in a single stage of 10 min at 680 °C and 100 MPa in tungsten carbide matrices of 8 or 15 mm diameter. These SPS processing conditions were selected after studies of the influence of pressure and temperature on the compactness of the composite electrode, in order to have sufficient compactness (86%) to ensure good ionic conductivity in the LAG. Applying the same parameters to a composite of formulation 45LVP/45LAG/10C leads to a compactness of only 70%. For equivalent composite electrode masses (50 mg) and 250 mg electrolyte, the pellet has a total thickness of 900 µm (Figure 4.5(b)).

The charge–discharge curve of a symmetric $Li_3V_2(PO_4)_3$/$Li_{1,5}Al_{0,5}Ge_{1,5}(PO_4)_3$/$Li_3V_2(PO_4)_3$ battery, denoted LVP/LAGP/LVP, with 15 mm diameter is shown in Figure 4.5(c). To compensate for the poor conductivity of LAGP at ambient temperatures (10^{-4} S/cm), the electrochemical tests were conducted at 80 °C above which the conductivity of LAG is at least 10^{-3} S/cm. As can be seen in Figure 4.5(c), the electrochemical performance of the all-solid-state battery is very similar to the equivalent battery using a liquid electrolyte (Figure 4.5(d)) [ABO 11].

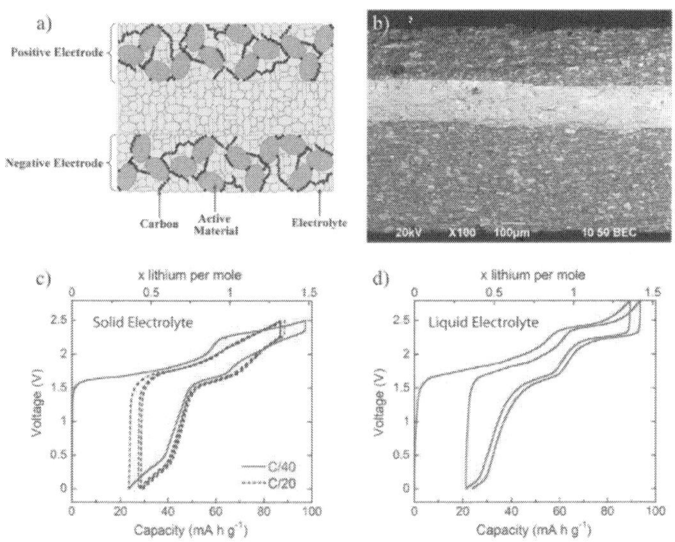

Figure 4.5. *a) Schematic of an all-solid-state battery. b) SEM/EDS cross-section of a LVP/LAGP/LVP solid-state cell (15 mm in diameter) assembled by SPS. c) Charge–discharge curves of the battery at 80 °C and a C/40 rate (1 Li^+ in 40 hr) and C/20. d) Galvanostatic charge/discharge curves of a symmetric LVP/LVP cell using a liquid electrolyte at C/10 rate and at ambient temperature [ABO 11]*

The charge–discharge curve of an LFP/LAG/LVP battery is shown in Figure 4.6(a). In the case of LFP, an LFP/LVP mass ratio of 0.7 is used to approach the maximum reversible capacity of the two electrodes (35 mg LFP, 50 mg LVP and 250 mg LAG). Cycling was conducted at 120 and 140 °C, temperatures which would be impossible for use with an organic electrolyte, under different regimes in order to test the limits of the batteries. The results show that the theoretical capacity of 80 mAh/g is obtained during cycles at 120 °C at a C/20 rate and that it decreases at this temperature if the rate doubles to C/10 (Figure 4.6(b)). After 30 cycles under different rate and temperature conditions, 80% of the initial capacity is obtained. This reduction in capacity is also observed for liquid electrolytes and is thought to be caused by the intrinsic properties of the negative electrode.

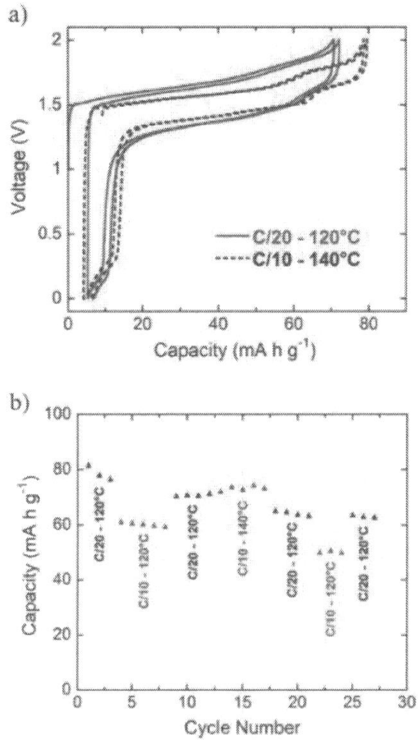

Figure 4.6. *a) Charge–discharge curves for an LFP/LAG/LVP battery (positive/ negative mass ratio = 0.7) at two different rates and temperatures for the potential window [0–2 V]. (b) Corresponding reversible capacity per gram of LVP, as a function of the number of cycles under different regimes and at different temperatures*

Impedance spectroscopy analysis was conducted in order to verify that the solid-to-solid contacts were maintained upon cycling (Figure 4.7). The loop observed at high frequency corresponds to the electrolyte and the loop at low frequency corresponds to the charge transfer at the active material/electrolyte interface and active material/electronic network interface as well as the ion transport through the composite electrode. The impedances of the electrolyte and the composite electrode decrease slightly but no significant change is observed in the cell's resistance, indicating that the ensemble of interfaces involved in the charge–discharge process remains stable.

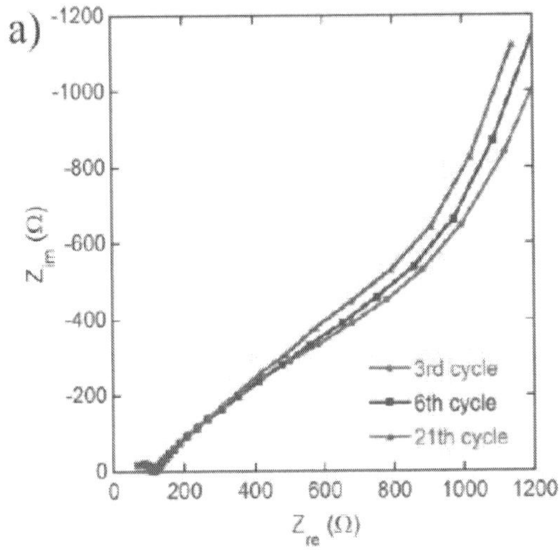

Figure 4.7. *Evolution of the impedance spectrum of an LFP/LAG/LVP battery at 120 °C in its discharged state*

The cells obtained after SPS sintering are self-supporting and exhibit good mechanical properties (Figure 4.8(a) and (b)) as well as well-defined electrode/electrolyte interfaces (Figure 4.6(c)). As a result of the thickness of these electrodes (300–800 μm), the LVP/LAG/LVP battery delivers a surface capacity of 1, 2 or 2.2 mAh/cm^2 depending on the selected cut-off voltage (1.75, 2.35 or 2.45 V, respectively) (Figure 4.8(d)), significantly greater than the few μAh/cm^2 delivered by a micro-battery. Values falling between 4 and 10 mAh/cm^2 were also obtained for other prototypes. Furthermore, these batteries are completely stable in air, unlike batteries using sulfur-based glass-ceramic electrolytes.

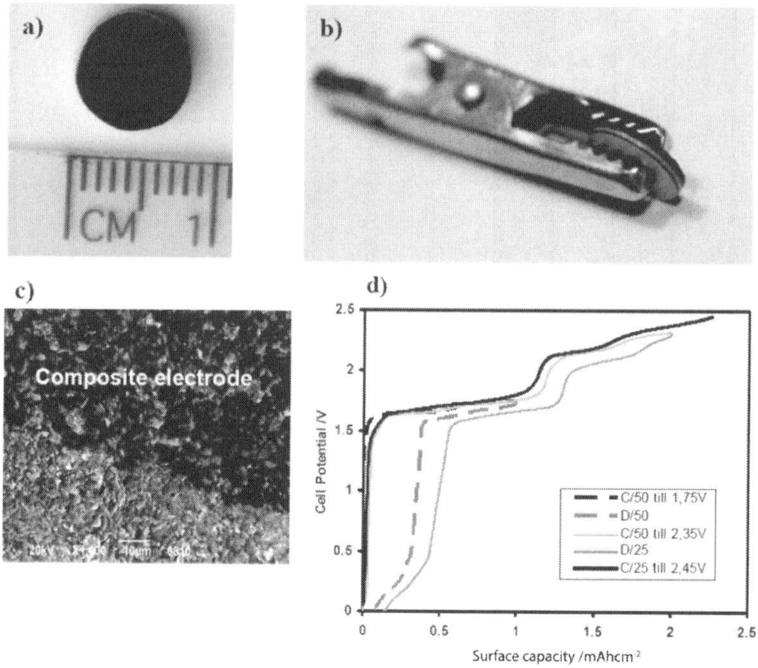

Figure 4.8. *(a, b) Photographs of a LVP/LAG/LVP battery obtained by SPS. (c) SEM image of the composite electrode/electrolyte interface in a solid-state battery (SPS densification parameters: 680 °C, 100 MPa). d) Charge/discharge profiles obtained at 80 °C and C/50 or C/25 for a symmetric LVP/LAG/LVP battery (25LVP/60LAG/15 Csp composite electrode)*

These promising results are unfortunately subject to at least two technological barriers:

– the first is the reactivity between the materials during thermal treatment. Most ceramic electrolytes require densification at very high temperature (≥ 650 °C) to achieve sufficient compactness for good ionic conductivity, which entails finding electrode materials that are likely to be completely inert during this treatment. Cathode materials like $LiCoO_2$ and $LiMn_{0.5}Ni_{0.5}O_2$ react badly with NASICON-type electrolytes at temperatures between 300 and 500 °C, and it has not yet been possible to assemble a solid-state battery;

– the second barrier concerns the electrode composition. As for batteries using glassy or glass-ceramic systems, the electrolyte here constitutes the skeleton of the battery, assuring the mechanical properties persist during cycling. A large fraction of the energy density is hence lost in the dead mass made up by the electrolyte.

To overcome these two problems, Ohta *et al.* proposed a provisional "monolithic micro-battery" type battery [OHT 12] with the garnet $Li_{6,75}La_3Zr_{1,75}Nb_{0,25}O_{12}$ as electrolyte, $LiCoO_2$ for the positive electrode and lithium for the negative electrode. To get around the problem posed by the garnet densification at 1200 °C, allowing it to reach 0.8×10^{-3} S/cm at 298 K, a 500 nm layer of $LiCoO_2$ was deposited by PLD on a 2 mm pellet of pre-sintered electrolyte. The assembly was subsequently heated to 600 °C. As shown in Figure 4.9, the battery exhibits excellent capacity retention of 125 mAh/g over 100 cycles at a current density of 3.5 μA/cm², or a rate of C/20 (1 Li^+ in 20 hr). Furthermore, no reactivity was observed between the $LiCoO_2$ and the garnet, even after 100 cycles, as demonstrated by SEM images of the electrolyte/electrode interface.

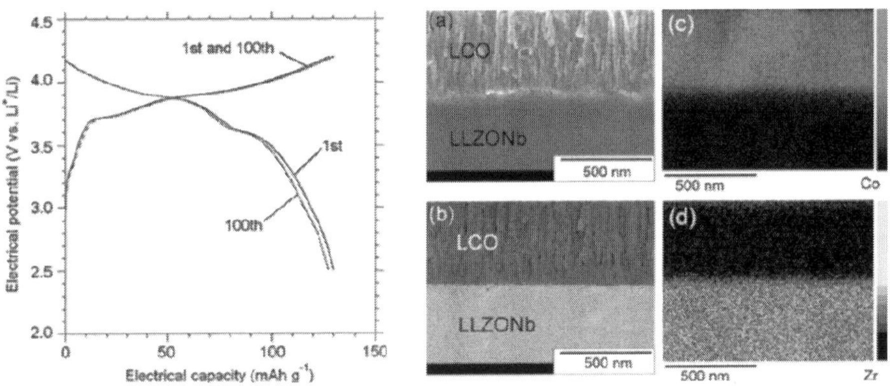

Figure 4.9. *Charge–discharge curve of the $LiCoO_2/Li_{6.75}La_3Zr_{1.75}Nb_{0.25}O_{12}/Li$ battery (left). SEM images of the electrode/electrolyte interface after 100 cycles (right) [OHT 12]*

Although the battery is very thick, it is difficult to refer to it as a monolithic battery because it only delivers a surface capacity of 32 μAh/cm². Nevertheless, the results obtained with this cell are encouraging for possible applications for these electrolytes. Furthermore, a $Cu_xV_2O_5/Li_7La_3Zr_2O_{12}/Li$ battery was also assembled following the same procedure [JIN 13].

4.3. Sodium "solid-state" battery technology

4.3.1. *Sodium-ion "solid-state" battery technology*

Sodium-based solid-state batteries were assembled by Lalère *et al.* in the same way as Li-ion batteries [LAL 15, LAL 14]. The material for the positive and negative electrodes, $Na_3V_2(PO_4)_3$, and the solid electrolyte $Na_3Zr_2Si_2PO_{12}$ were selected after consideration and testing of a large number of target materials.

The thermal stability of different electrolyte/active material pairs was investigated. The SPS sintering temperature determined for $Na_3Zr_2Si_2PO_{12}$ was 900 °C. This is much higher than the synthesis temperatures for most materials and would be expected to correspond to high reactivity. However, as shown by the XRD spectra of mixtures sintered by SPS at different temperatures in Figure 4.10, the mixture $Na_3V_2(PO_4)_3/Na_3Zr_2Si_2PO_{12}$/Csp (25/60/15) exhibits only weak chemical reactivity, even at 900 °C. The spectra for mixtures at 800 and 900 °C are superpositions of the spectra for $Na_3V_2(PO_4)_3$ and $Na_3Zr_2Si_2PO_{12}$, except for a few small impurities of unknown nature that produce the weak reactivity.

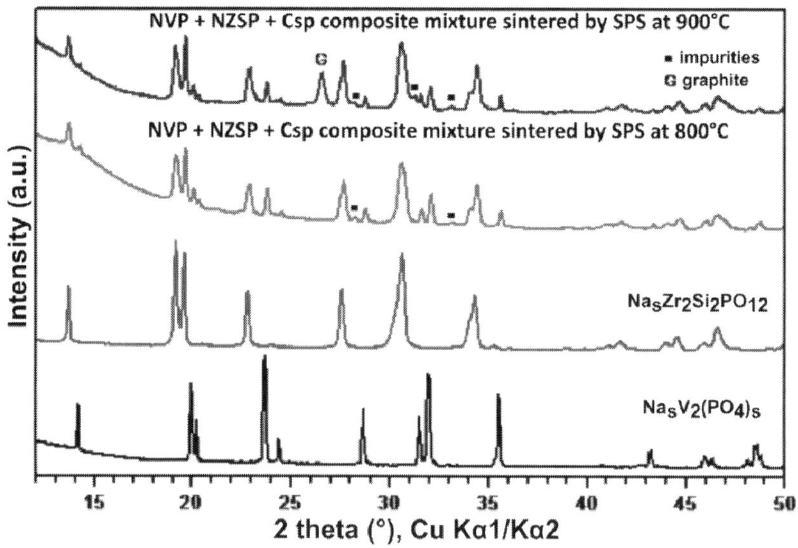

Figure 4.10. *X-ray diffraction patterns of $Na_3V_2(PO_4)_3/Na_3Zr_2Si_2PO_{12}$/Csp (25/60/15) mixtures sintered at 800 °C (green) and 900 °C (blue) compared to reference patterns for the constituent materials (black and red). For a color version of this figure, see www.iste.co.uk/viallet/batteries.zip*

The composite mixture is identical for the two electrodes and is obtained by simple mortar grinding of $Na_3V_2(PO_4)_3$ (25% by mass), $Na_3Zr_2Si_2PO_{12}$ (60%) and *Carbon Super P®* (15%). The particle sizes were of the order of 10, 50 and 0.01 μm for NVP, NZSP and Csp, respectively. NVP is the active material in this mixture, but also acts as a mechanical binder because the SPS fabrication temperature is greater than its sintering temperature. The NZSP ensures ionic conductivity and is the main mechanical binder. Finally, the carbon provides the electronic conductivity and also serves as the dispersant.

Double the quantity of the composite mixture was introduced into the negative electrode in order to balance the capacities of the two electrodes. Next, the matrix was placed in the device with the same orientation, i.e. the conventional setup with the negative electrode on the bottom and the positive electrode on the top. Hence, the direction of the continuous current during sintering is always the same with respect to the pellet for all experiments. The sintering was carried out in the same way as for the sintering of the electrolyte by itself. The temperature was held at 900 °C and the pressure held at 100 MPa for 10 min before the sample was left to cool. After SPS densification, the battery was removed from the mold, the graphite layer was removed and the pellet was photographed (Figure 4.11(a)). The three different layers of the battery are clearly visible, with – from top to bottom – the positive composite electrode, the solid electrolyte and the negative electrode. After polishing, the battery was observed under a scanning electron microscope (Figure 4.11(b)). The three layers have thicknesses of 130, 130 and 300 μm for the positive composite electrode, the solid electrolyte and the negative composite electrode, respectively. The difference in thickness between the two electrodes results from the difference in mass initially introduced to balance the capacities. The negative electrode capacity is thus ~1.15 times greater than the positive electrode's (117.6 mAh/g of NVP at the positive electrode (130 μm) compared to 58.8 mAh/g of NVP at the negative electrode (300 μm). The positive/negative capacity ratio can be expressed as (130×117.6) / $(300 \times 58.8) = 1.15$).

The interfaces are remarkably well-defined between each layer, but also within the composite electrodes, in which a good distribution of each base material can be seen.

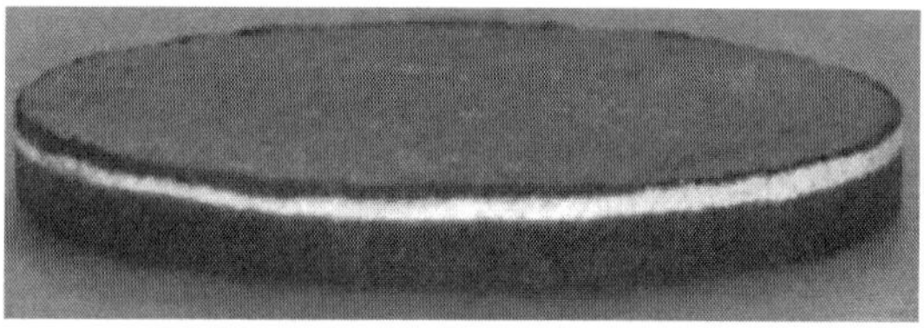

a)

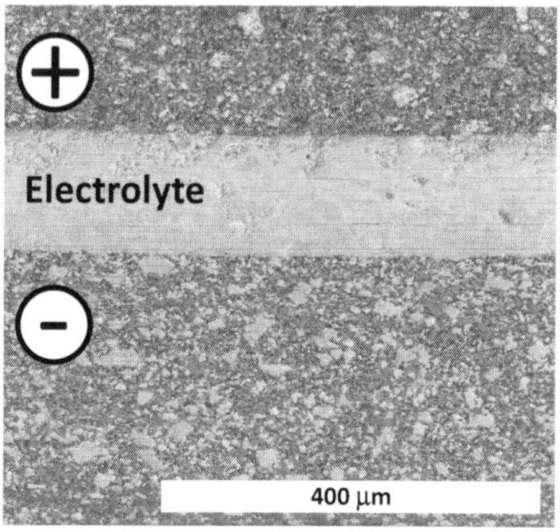

b)

Figure 4.11. *a) Photograph of a $Na_3V_2(PO_4)_3$: $Na_3Zr_2Si_2PO_{12}$: Csp (25:60:15)/$Na_3Zr_2Si_2PO_{12}$/$Na_3V_2(PO_4)_3$: $Na_3Zr_2Si_2PO_{12}$: Csp (25:60:15) battery and (b) corresponding SEM image in which NZSP appears as light gray, NVP as dark gray and Csp as black*

Scanning electron microscopy with energy dispersive X-ray spectroscopy also allowed the elements at the surface of the battery to be mapped, demonstrating clearly that no interdiffusion took place during the SPS fabrication of the battery.

After being dried under vacuum, the battery was placed into the high-temperature electrochemical cell (Figure 3.17) under an argon atmosphere and heated to 200 °C for galvanostatic analysis ($\sigma_{(NZSP)200°C} > 10^{-2}$ S/cm). Figure 4.12 shows (a) the battery behavior for cycling at different current rates, (c) the corresponding capacity retention and (d) the *post-mortem* SEM analysis. Figure 4.12(b) shows the first cycles at 200, 250 and 300 °C, which all exhibit an "extra-capacity".

The galavanostatic cycling of the NASICON "solid-state" battery for a $1e^-/5h$ rate is shown in blue in Figure 4.12(a). The current rate is calculated with respect to the positive electrode, i.e. 1e⁻/xh corresponds to charging the battery in 2xh. The electrochemical curve profile is very close to the reference liquid electrolyte battery's, with weak polarization, 85% of the theoretical capacity and a plateau at a predictable voltage of ~1.7 V.

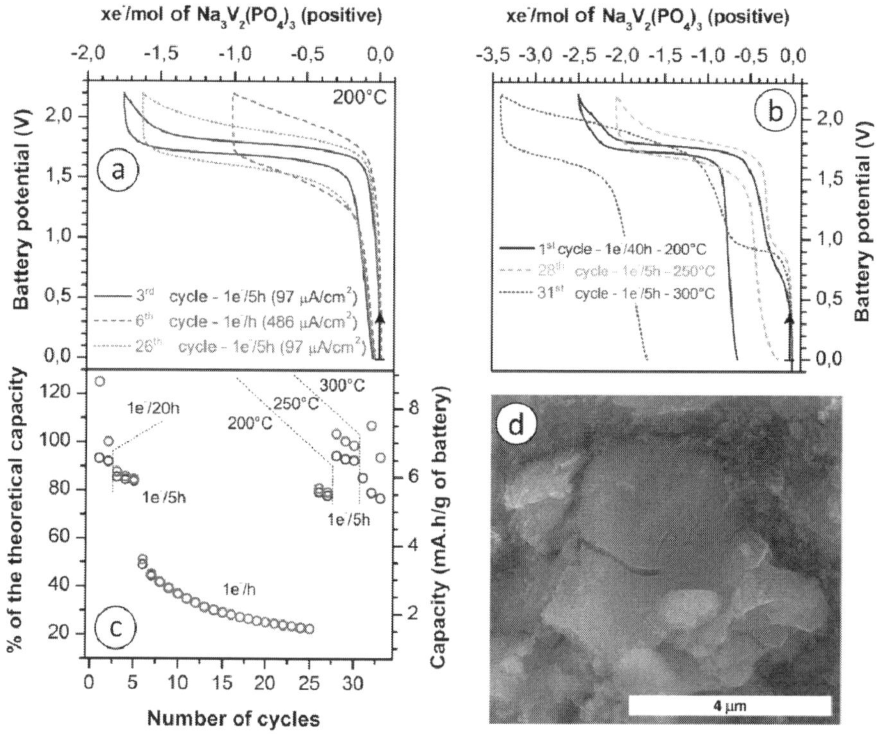

Figure 4.12. *(a, b) Galvanostatic cycling of a $Na_3V_2(PO_4)_3$: $Na_3Zr_2Si_2PO_{12}$: Csp (25:60:15)/$Na_3Zr_2Si_2PO_{12}$/$Na_3V_2(PO_4)_3$: $Na_3Zr_2Si_2PO_{12}$:Csp (25:60:15) battery for different current rates at 200, 250 and 300 °C, (c) evolution of the corresponding capacity and (d) SEM image of a $Na_3V_2(PO_4)_3$ grain in the positive electrode after 33 cycles. For a color version of this figure, see www.iste.co.uk/viallet/batteries.zip*

Moving to five times faster kinetics induces a significant increase in the polarization and changes the curve profile, which results in a drop in capacity. This capacity does not reach more than 25% of the theoretical capacity after 20 cycles at high regime. Nevertheless, when the current regime is decreased back to 1e⁻/5h, the capacity returns to 80%, albeit with a polarization that is still high. This polarization is probably due to damage (mechanical, for example) sustained by the electrodes during cycling at high regimes and/or at 300 °C. Overly rapid changes in volume during the insertion/extraction of sodium could cause cracks in the $Na_3V_2(PO_4)_3$ grains, as demonstrated by the SEM image in Figure 4.12(d).

An extra capacity of variable prominence, could be observed (or not) only on the charge curves according to the cycling conditions (Figure 4.12(b)). From the first charge of the battery at a rate of $1e^-$/20h and a temperature of 200 °C, an irreversible electrochemical activity counting for $0.5e^-$ appears at a voltage of ~0.7 V. On the second cycle, the phenomenon is still present but weakened, whereas on the third cycle, at a four times faster regime, it is absent. It reappears once the temperature reaches 250 °C, however, even for a regime of $1e^-$/5h. A decrease in polarization can be observed with this increase of 50 °C between the 26th and the 28th cycle. This is explained by the improvement in material conductivities, which, between 200 and 250 °C, increase from 3×10^{-2} to 5×10^{-2} S/cm and from 9×10^{-3} to 2×10^{-2} S/cm for $Na_3Zr_2Si_2PO_{12}$ and $Na_3V_2(PO_4)_3$ respectively. Finally, at 300 °C, still with the same regime, the charge capacity is around 170% of the theoretical capacity for the 31st cycle (not visible in the figure) but is only 50% reversible.

This study demonstrated not only the feasibility of the sodium system, but also encourages more detailed work to develop comprehension and optimization. Indeed, the large particle size, the low percentage of active material and the slight electrode thicknesses are parameters ripe for improvement. In order to optimize the electrochemical performance of this NASICON all-solid-state battery, the influence of the galvanostatic cycling, SPS fabrication and powder preparation conditions were studied.

The powder preparation, SPS fabrication conditions and electrochemical cycling conditions all affect the battery performance. Figure 4.13 repeats the study of galvanostatic cycling conditions, but this time for a battery with nanoscale grains of $Na_3V_2(PO_4)_3$.

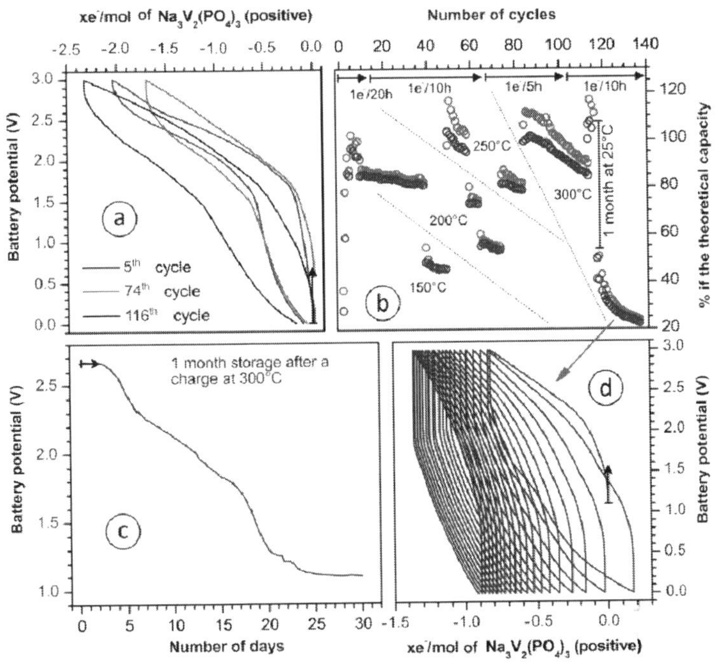

Figure 4.13. *a) Galvanostatic cycling of a NASICON all-solid-state battery at 200, 250 and 300 °C. b) Corresponding capacity retention. c) Evolution of the battery voltage after a month's storage at 25 °C. d) Galvanostatic cycling at 300 °C after a month's storage at 25 °C. For a color version of this figure, see www.iste.co.uk/viallet/batteries.zip*

Panels (a) and (b) show that the battery can withstand more than 100 cycles despite significant variations in current rate and temperature. For the same current rate, increasing the temperature by 50 °C allows the capacity to improve by 20–30%. At constant temperature, increasing the current regime by a factor of two induces a drop in capacity of roughly 20%. The current regime and temperature thus have opposite effects on the capacity retention such that an increase in the regime from $1e^-/10h$ to $1e^-/5h$ is compensated by an increase in the temperature from 200 to 250 °C. Too high a temperature negatively impacts the capacity retention over time, as demonstrated by cycling at 300 °C.

Independent of the electrochemical profile, the energy density can be improved by increasing the total percentage of active material in the battery, i.e. by increasing the electrode thicknesses and by increasing the amount of active material in the electrodes. However, both efforts risk decreasing the power density and capacity retention. Ultimately, the operating temperature, the current regime and the cut-off voltage must all be co-ordinated, along with the battery type, to guarantee good electrochemical properties for the intended use.

The design of NASICON solid-state batteries requires compromises between fabrication and operation. However, whatever compromises are made, the performances of this battery system tend toward the same values of 10–12 and 0.2–4 W/kg, as shown in Figure 4.14. Figure 4.14 also shows the performances of NASICON solid-state batteries compared to other battery technologies. These performances only take into account the pellet, without any support, protection or heating system, connected or otherwise.

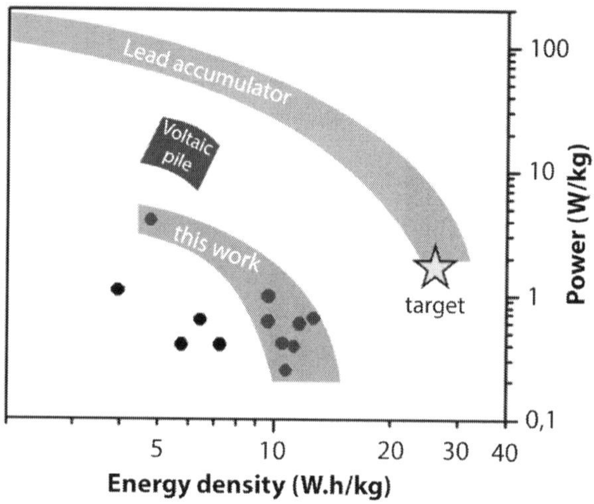

Figure 4.14. *Energy and power densities of NASICON solid-state batteries compared to other battery technologies. The star demarks what can reasonably hoped to be achieved*

In order to critically consider the experimentally obtained energy densities and determine the performance that NASICON "solid-state" batteries could attain, Lalère *et al.* sought to calculate the evolution of the energy density as a function of the active material content in the electrodes, and of the electrode thicknesses. To approximate real conditions, a symmetrical NVP/NZSP/Csp battery was considered, taking the form of a cylinder with 10 mm diameter, with an average compactness of 80% and with an 100 μm electrolytic layer and a carbon content of 15% in the composite electrodes (Figure 4.15).

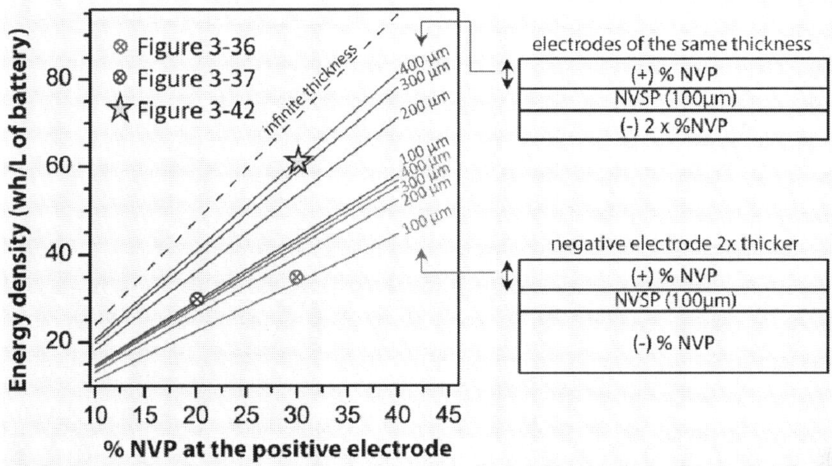

Figure 4.15. *Theoretical energy densities of a NASICON all-solid-state battery as a function of the positive electrode thickness and the percentage of $Na_3V_2(PO_4)_3$ in this electrode for (in red) the same percentage in the two electrodes and (in blue) double the percentage at the negative electrode. The black dotted line represents the energy density limit that this battery system can achieve. The red and blue crosses correspond to batteries produced in the laboratory. The star demarks what can reasonably hoped to be achieved. For a color version of this figure, see www.iste.co. uk/viallet/batteries.zip.*

The red lines delineate the evolution of energy density (by volume) when the battery electrodes' capacities are balanced by differences in the electrode masses/thicknesses, with identical active material content. The blue traces show the evolution in energy density when the capacities are balanced by different active material content in the electrodes, with identical electrode masses/thicknesses.

Whatever capacity-balancing method is used, the energy density tends toward a limiting value that increases with thickness. The greater the thickness of the electrodes, the smaller the relative thickness of the electrolytic layer and hence the lower the proportion of inactive material. Thus for electrodes infinitely thicker than the middle layer, the energy density only depends on the NVP content in the electrodes (black dotted line). Whatever the NVP content, the thickness of the positive electrode must not exceed 300–400 µm because the energy density improvement for thicknesses above this is slight.

4.3.2. *Sodium-sulfur "all-solid-state" battery technology*

Recently, Song *et al.* have also assembled a solid-state battery exhibiting impressive performance with a Nasicon-type oxide, $Na_{3.1}Zr_{1.95}Mg_{0.05}Si_2PO_{12}$ [SON 16]. With the aim of demonstrating the advantages of solid-state technology, the performance of a solid-state cell was compared to that of a cell with liquid electrolyte. The all-solid-state battery, of the button cell type, is the combination of a sulfur-carbon-polymer hybrid positive electrode, a 1-mm-thick ceramic pellet of the solid electrolyte $Na_{3.1}Zr_{1.95}Mg_{0.05}Si_2PO_{12}$ and metallic sodium for the negative electrode, the ceramic pellet also serving as the separator. The other cell comprises the same electrodes, but separated by a glass microfiber, the liquid electrolyte being a 1 M $NaClO_4$/EC+PC mixture (1:1 by mass). The cell with liquid electrolyte (open symbols) has an initial capacity in the region of 180 mA/g, but which decreases to around 64 mA/g after 100 cycles at a 1C rate with a capacity retention of 36% (Figure 4.16(a)). The initial capacity for the solid-state cell (full symbols) is slightly lower at 170 mA/g, but exhibits a higher value of 150 mA/g after 100 cycles, with a capacity retention of 88%, indicating better cyclability compared to the Na–S cell employing a liquid electrolyte. For a 5C regime, while the initial capacity of the cell with liquid electrolyte is higher at 84 mA/g, it tends toward the same value as for the solid-state cell, 60 mA/g, after 100 cycles.

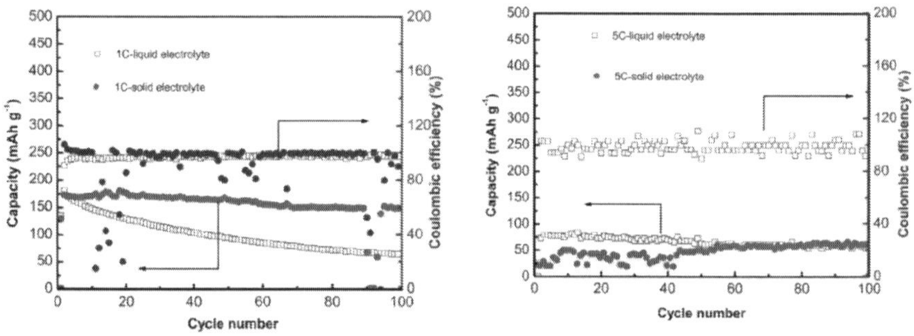

a) b)

Figure 4.16. *Electrochemical performance of the two cells, with solid and liquid electrolyte, at a 1C rate (a) and a 5C rate (b) [SON 16]. For a color version of this figure, see www.iste.co.uk/viallet/batteries.zip*

These results demonstrate that solid-state Na–S cells can exhibit better performance than analogs using a liquid electrolyte, the solid electrolyte enabling a reduction in the dissolution and migration of sodium polysulfides in the electrolyte.

5

$LiBH_4$ Electrolyte and Polymer Battery Technology

5.1. "All-solid-state" battery technology: $LiBH_4$ electrolyte

In 2013, Takahashi *et al.* reported the first "all-solid-state" battery using a borohydride electrolyte [TAK 13]. The battery was cycled at 120 °C, the temperature at which $LiBH_4$ exhibits sufficient conductivity, and used an electrode composed of $LiCoO_2$ and $LiBH_4$ with 10 to 1 mass ratio. The cell exhibited an initial capacity of 157 mAh/g, which very rapidly dropped to 18 mAh/g on the 10th cycle as a result of strong reactivity between the $Li_{1-x}CoO_2$ and the $LiBH_4$. This reactivity was observed by Raman spectroscopy and allowed the following reactivity equation to be established:

$$Li_{0.5}CoO_2 + 1/8\ LiBH_4 \rightarrow 1/4\ Co_3O_4 + 1/4\ CoO(OH) + 1/8\ LiBO_2 + 1/4\ Li_2O + 1/8\ H_2$$

It was found that this reactivity could be limited by fabricating a thin-film electrode composed of a layer of $LiCoO_2$ and a layer of Li_3PO_4 to protect it from the $LiBH_4$. The charge-discharge curve for this battery is shown in Figure 5.1.

In 2015, Unemoto *et al.* assembled an all-solid-state battery with $LiBH_4$ for the solid electrolyte, TiS_2 for the active material at the positive electrode and lithium at the negative electrode [UNE 15]. This achieved very high performance over more than 300 cycles at 120 °C at a 0.2C rate, with a capacity of 250 mAh/g on the second cycle and more than 180 mAh/g on the 300th cycle with a Coulombic efficiency close to 100%. The slight decrease in performance could have been linked to chemical/electrochemical oxidation of the $LiBH_4$ at the TiS_2 surface at the operating temperature and/or during the initial charge. During this oxidation reaction, an additional phase, probably incorporating $Li_2B_{12}H_{12}$, precipitated at the interface between the $LiBH_4$ and the TiS_2. $Li_2B_{12}H_{12}$ has a lithium ionic conductivity of around 5×10^{-5} S/cm at 120 °C and acts as a protective layer between the solid electrolyte and the TiS_2 active material.

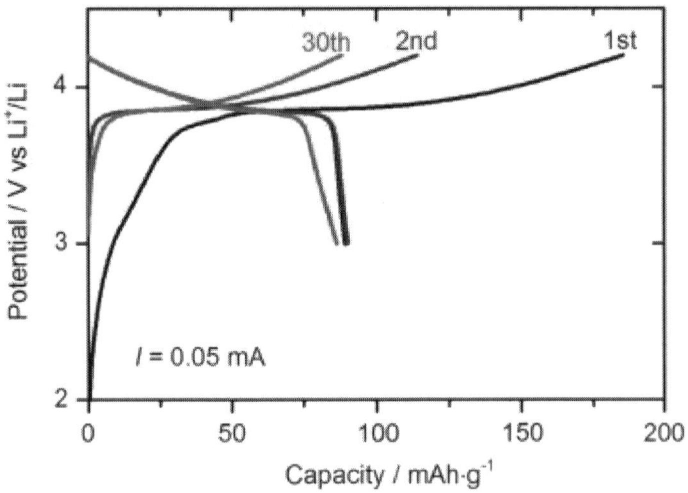

Figure 5.1. *Charge–discharge curve for a battery composed of LiCoO2/Li3PO4/LiBH4/Li, at 0.05 mA and 120 °C [TAK 13]. For a color version of this figure, see www.iste.co.uk/viallet/batteries.zip*

More recently, solid-state lithium-sulfur batteries were assembled from $LiBH_4$ confined in a mesoporous silica matrix [DAS 16]. This nano-confinement enables an increase in the ionic conductivity of $LiBH_4$ to 0.1 mS/cm at ambient temperature, with a cationic transport number of 0.96. This electrolyte exhibits excellent stability compared to lithium. A capacity of 1220 mAh/g after 40 cycles was achieved at a temperature of 55 °C and a rate of 0.03C.

5.2. "Solid-state" polymer battery technology

Rechargeable lithium-polymer batteries represent an excellent source of electrochemical energy due to their high energy and/or power density, their good cyclability, their reliability and their safety [ARA 97, BEA 95, KAL 96, SCR 95]. A rechargeable lithium polymer battery is generally formed of a metallic lithium anode, a polymer electrolyte separator and a metal oxide cathode capable of reversibly intercalating and liberating (de-intercalating) lithium ions.

Complexes of lithium salts and polyethylene oxide, PEO-Lix, are very promising candidates for electrolyte separators [LIG 93, VIN 93]. These polymer electrolytes exhibit attractive ionic conductivities ($\sigma > 10^{-4}$ S/cm) above 60 °C, typically in the range of 80–100 °C [ARM 89, GRA 97, GRA 00, LIG 93]. This high temperature limits their use and is inconvenient for application in electronics, but can be advantageous for many other applications such as for electric vehicles, energy storage and charge equalization systems. Consequently, considerable R&D effort has been dedicated to the formulation and development of PEO-based electrolytes capable of combining high conductivity with superior interfacial stability for a metallic lithium anode, as well as good mechanical properties [APP 00a, APP 01a, APP 00b, APP 01b, APP 00c, BOR 95, CAP 91, CRO 98, PRO 98, WIE 89].

Another crucial aspect for the development of a lithium-polymer battery is the selection of an appropriate cathode. The most common cathode materials, such as $LiCoO_2$, $LiNiO_2$, $LiCo_xNi_{1-x}O_2$, and $LiMn_2O_4$, display a loss in capacity at 70 °C, the temperature at which polymer electrolytes are used, mostly due to reactions with the electrolyte or phase transitions [VIN 93]. Thermostable materials that could be used in their place, such as vanadium oxides like V_2O_5 or V_3O_8 [VIN 93, SCA 93, DEL 95], are relatively expensive and pose problems with toxicity.

6

Markets

6.1. Solid electrolytes

6.1.1. *Ohara*

Ohara Corporation has developed lithium ion conducting glass-ceramics intended for use as solid-state electrolytes or separators in next-generation batteries or electrochemical devices[1]. Ohara offers two types of membranes: LICGC™ AG-01 membranes obtained via melting and polishing, based on $Li_2O–Al_2O_3–SiO_2–P_2O_5–TiO_2–GeO_2$ with a conductivity of 1×10^{-4} S/cm (Figure 6.1), and LICGC™ SP-01 membranes obtained via doctor blading and sintering at 25 °C, based on $Li_2O–Al_2O_3–SiO_2–P_2O_5–TiO_2$ with conductivity between 2.5 and 4×10^{-4}S/cm at 25 °C.

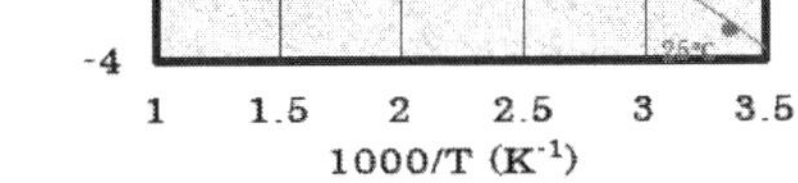

Figure 6.1. *The ionic conductivity of LICGC™ (AG-01 membranes) plotted in Arrhenius coordinates*

1 Retrieved September 27, 2015, from http://www.oharacorp.com/lic-gc.html.

The technical properties of LICGC™ can be found online and in a downloadable pdf presentation [NAK 10]. The ion transport number is 1.

This glass-ceramic exhibits excellent chemical resistance. Water and weak acids have little effect on the conductivity. It is stable in air and water (Figure 6.2(a)), and is non-flammable (Figure 6.2(b)).

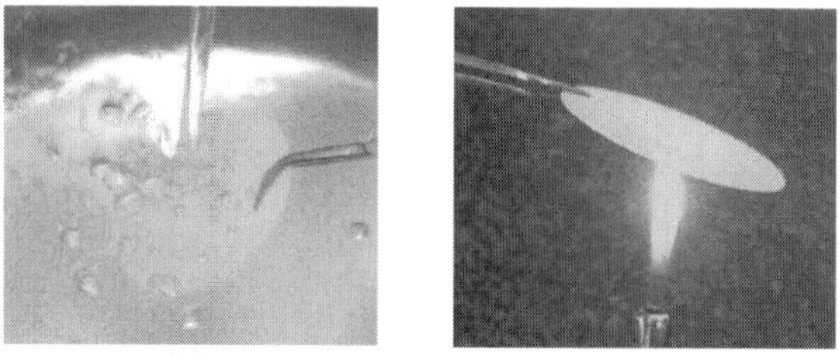

(a) (b)

Figure 6.2. *Demonstration of (a) the water stability and (b) non-flammability of the LICGC™ membrane*

It is thermally stable up to 600 °C (Figure 6.3).

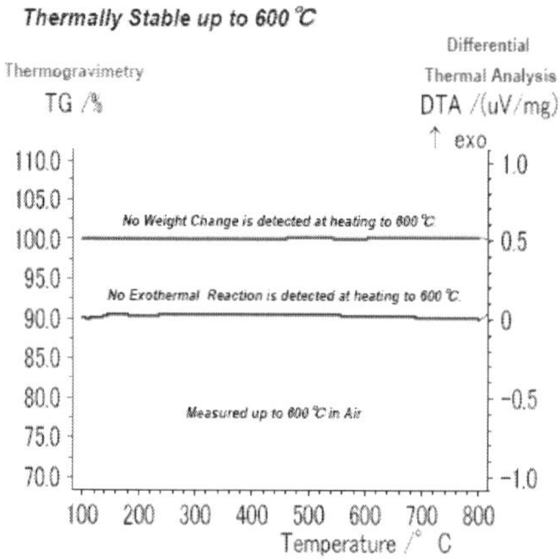

Figure 6.3. *Thermogravimetric (red*) and differential thermal analysis (blue) of the LICGC™ conducting membrane. For a color version of this figure, see www.iste.co.uk/viallet/batteries.zip*

The membranes are available in a variety of sizes (2.54cm (1″) sided squares with 150 µm thickness, 5.08 cm (2″) sided squares with 200 µm thickness or 5.08 cm diameter discs with 250 μm thickness). The company can supply discs with diameters up to 15.24 cm (6″). The melting fabrication process (Figure 6.4) allows these dimensions and quantities to be obtained.

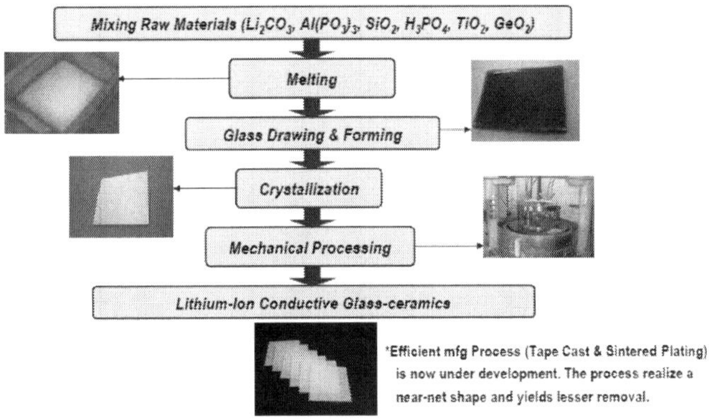

Figure 6.4. *Fabrication steps for the AG-01 membrane, including raw material mixing, melting, forming and crystallization steps*

The AG-01 glass-ceramic is composed of two NASICON-type crystalline phases with formulae $Li_{1+x}Al_{lx}Ge_yTi_{2-x-y}P_3O_{12}$ and $Li_{1+x+3z}Al_x$ $(Ge,Ti)_{2-x}$ $(Si_zPO_4)_3$, combined with $AlPO_4$ impurities (Figure 6.5). These membranes have been successfully tested in Li-air and Li-seawater devices developed by POLYPLUS [NAK 10] and in Li-sulfur devices (Eurolis European Project [VIZ 14]). However, this leaves the issue of the membranes' electrochemical stability, which is not addressed: various applications show that the membranes are not stable alongside the lithium metal and make use of a protective layer (the basis of a large number of patents).

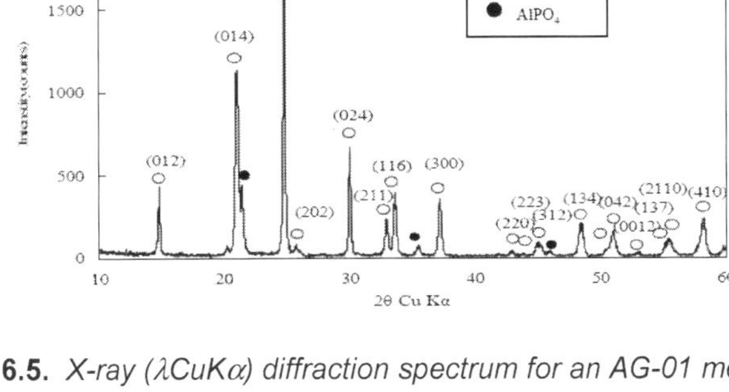

Figure 6.5. *X-ray (λCuKα) diffraction spectrum for an AG-01 membrane*

Ceramic membranes have been used in Li-sulfur batteries to prevent polysulfides, most of which are soluble, from acting as a redox shuttle inducing serious self-discharge. The membrane, a solid electrolyte, can effectively separate the electrodes and ensures good ionic conductivity. A battery has been assembled with the AG-01 membrane in addition to a fiberglass separator to avoid contact between the ceramic membrane and the lithium in the anode. A porous sulfur composite electrode soaked with liquid electrolyte is in direct contact with the ceramic membrane (Figure 6.6(a)). This configuration demonstrated excellent stability over more than 350 deep charge/discharge cycles (Figure 6.6(b)), at a rate of C/20 and an average Coulombic efficiency above 99.95%.

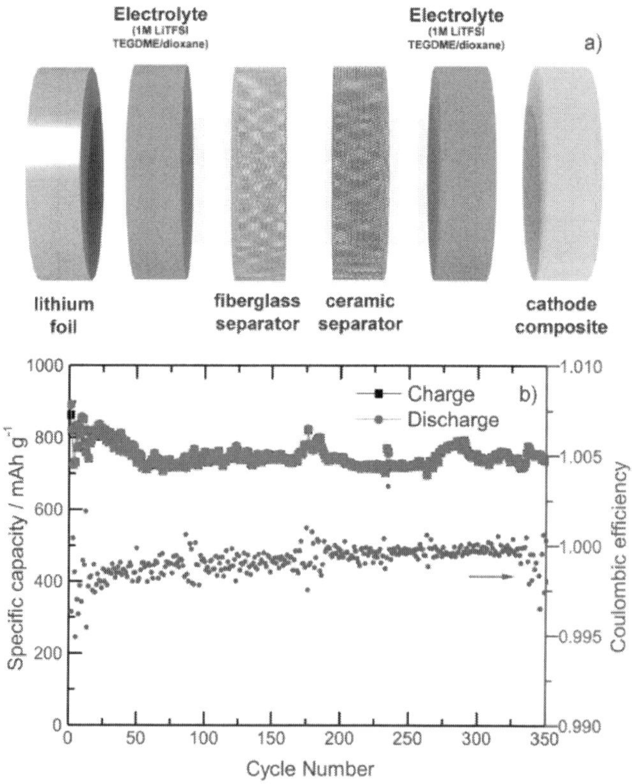

Figure 6.6. *(a) Configuration of the Li–S battery with ceramic membrane and (b) measured capacity (left-hand axis) and Coulombic efficiency (right-hand axis) over more than 350 cycles, for a C/20 cycling regime at 25 °C [VIZ 14]. For a color version of this figure, see www.iste.co.uk/viallet/batteries.zip*

6.1.2. *NEI*

In May 2013, NEI Corp. announced the commercialization of $Li_{10}SnP_2S_{12}$ as a solid-state electrolyte for solid-state battery applications [NEI press release]. Work on this phase started shortly beforehand at LRCS and at ICMCB (see Tarhouchi's thesis [TAR 15]) in order to answer longstanding questions surrounding the structure and material properties of the LGPS structure. This study thus set about characterizing these materials, focusing on those based on tin and silicon, in order to better understand their properties.

Three different batches of $Li_{10}SnP_2S_{12}$ were characterized during the thesis, labeled NEI1, NEI2 and NEI3. Although they were all reported to be relatively pure (> 98%), the three materials received turned out to be quite differently structured.

Rietveld refinements of the diffractograms of the highly crystalline NEI1 and NEI3 samples clearly showed that they were mostly composed of a $Li_{10+\delta}Sn_{1+\delta}P_{2-\delta}S_{12}$ phase with LGPS structure. This structure is different from the thio-LiSICON analog form proposed for the germanium and silicon systems before the publication by Kamaya *et al.* of the LGPS structure [Kanno 2011]. Rietveld refinements also demonstrated the presence of Li_2SnS_3 impurities in NEI1 and Li_4SnS_4 impurities in NEI3.

Other impurities, which were not detected by XRD, could be detected by NMR of ^{31}P in the two materials: an oxide with a signal close to that of γ-Li_3PO_4 and a β-Li_3PS_4 phase. These two metastable phases display an arrangement of PX_4 (X=O, S) tetrahedrons analogous to the LGPS structure. Two-dimensional NMR measurements confirmed that these signals belonged to phosphorous in other phases, and hence to impurities.

6.2. Solid-state batteries

Figure 6.7 shows the energy densities of current battery technologies. It can immediately be seen that the traditional lead/acid battery, which is very heavy and bulky, stores little energy.

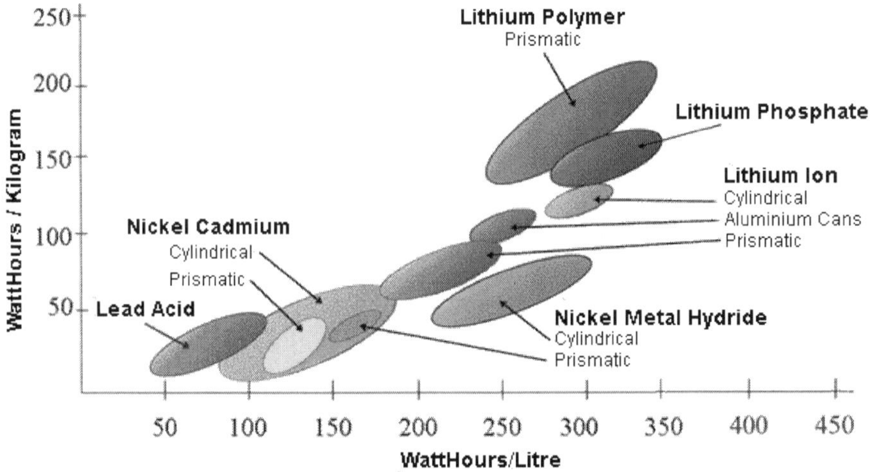

Figure 6.7. *Energy densities of current battery technologies[2]. For a color version of this figure, see www.iste.co.uk/viallet/batteries.zip*

It is worth examining the capacities of batteries fitted to electric cars on the market today[3].

The Renault ZOE is equipped with a lithium-ion battery made by LG Chem. The usable capacity is 22 kWh (24 kWh total capacity), with a weight of 290 kg and an unknown volume for a range of 150 km. A new lithium-ion battery of around 4 kWh, supplied by LG Chem, drives the 2017 electric Renault ZOE. Thus, Renault aims to double the range, which allows the car to drive 300 km under real conditions and 400 km on the European test cycle[4].

The lithium-ion battery in the Nissan LEAF, with a capacity of 24 kWh, offers a range of 199 km* (*New European Driving Cycle – NEDC). It is 95% recyclable. Nissan will have a battery with double the energetic capacity, at close to 50 kWh, for its next electric Leaf from 2017. However, the source cited by "Les Echos" did not specify if the Japanese group will

2 Retrieved September 25, 2015, from http://www.mpoweruk.com/chemistries.htm,
3 Retrieved September 25, 2015, from http://blog.erios.org/index.php?post/2013/11/18/Stockage-de-l-%C3%A9lectricit%C3%A9%3A-les-batteries-du-futur
4 Retrieved September 25, 2015, from http://bfmbusiness.bfmtv.com/entreprise/renault-l-autonomie-de-la-zoe-va-passer-a-300-km-859555.html

continue to use its joint venture with NEC for supplies, or instead the Korean LG (as Renault does for the ZOE), a sensitive subject in the alliance between the two automobile constructors [BFM 15].

The BMW i3 is sold with a lithium-ion battery made by Samsung SDI. The total capacity is 22 kWh for a weight of 230 kg and an unknown volume, with an estimated real range of 150 km.

The Tesla Model S is sold with a lithium-ion battery made by Panasonic, based on cells with 18650 A form factor. The total capacity is 85 kWh for a weight of 544 kg, and an estimated real range of 400 km.

If we calculate the specific energy of each model, taking account of the total capacity of each battery, we obtain 82Wh/kg for the Renault ZOE, 95 Wh/kg for the BMW I3 and 156 Wh/kg for the Tesla Model S.

Solid-state technology is already on the market. The Bolloré BlueCar has an LMP battery (Figure 6.8[5]) developed by BatHium Canada Inc. and BatScap, subsidiaries of the Bolloré group (a business 80% owned by Bolloré and 20% by EDF).

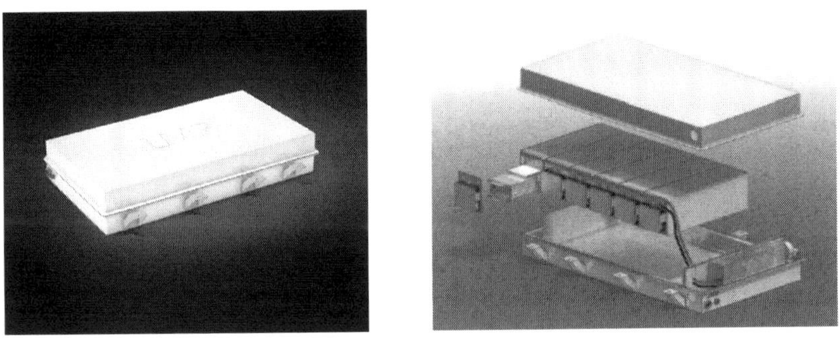

(a) (b)

Figure 6.8. *LMP® battery designed by Batscap and fitted in bluecars*

The technology used is called "Lithium Metal Polymer". The anode is hence metallic lithium, the solid electrolyte is PEO-LITFSI (poly(ethylene oxide) (PEO) doped with a lithium salt, bis(trifluoromethanesulfonyl)imide

5 Retrieved September 25, 2015, from http://www.bluecar.fr/les-batteries-lmp-lithium-metal-polymere

(LiTFSI), and the cathode is $LiFePO_4$[6]. The total capacity is 30 kWh for a weight of 300 kg and a volume of 300 L, with an estimated real-world range of 200 km. The specific energy is hence 100 Wh/kg.

This LMP® battery has powered Bluecars for the car-hire service Autolib' since December 2011. The official bluecar website (Figure 6.8) states that "No matter the weather conditions encountered, the range of users, the driving conditions – both urban and sub-urban – or even after fire testing, nothing has affected battery performance. Through extensive user-testing, the LMP battery has demonstrated its reliability, its technological performance and its durability. The fleet has already racked up close to 100,000,000 km since 5 December 2011". Furthermore, "As a result of its 'solid-state' design, it is safer than other next-generation battery technologies. The LMP® battery is particularly robust to all weather conditions and hence offers true user safety. Made entirely from non-polluting materials, it does not pose any threat to the environment. At the end of its life, all of its components will be recycled or repurposed".

Bluecars recharge in a dozen hours using a standard plug socket limited to 10 A delivered via a Mode 2 charging cable equipped with a domestic plug that comes with the car. To make use of the full charging power and fully recharge in 8hr at 16 A, the customer must buy a dedicated wall-box sold by Bolloré (for €995).

The research and development of this technology was organized in Canada in 1979, starting with work by the HydroQuebec research institute, in association with other international partners including the Centre National de la Recherche Scientifique (CNRS) in France, the YUASA Corporation in Japan, the United States Advanced Battery Consortium (USABC) and the 3M Corporation. This coalition later signed up as partners to the SEVE project to address the objectives of the Kyoto protocol. This protocol represents a truly global challenge, with the transport sector implicated in 40% of greenhouse gas emissions. From 1995 onward, the Bolloré group started to develop its own Lithium Metal Polymer battery technology, using different processes, in its research and development laboratories at Ergué Gabéric, in Finistère (Brittany, France). It would go on to create the BATSCAP company, dedicated to this activity, in 2004.

6 Retrieved September 25, 2015, from http://www.bathium.com/

By taking control of the assets belonging to the company Avestor, owned by HydroQuebec and Anadarko, in 2007, the Bolloré group confirmed its dominance of the LMP battery technology. On January 24, 2008, the group launched its first line of supercapacitors and very quickly built up a solid reputation in this area of expertise, by buying out Avestor, which became Bathium and Batscap. The Bolloré Group also confirmed its legitimacy in LMP technology and became a major international actor in the energy sector. The high performance Lithium Metal Polymer electric battery, through its combination of technologies, complex production processes and integration of 3M LiTFSI lithium salts, endows Bluecar® with its performance and gives the Autolib' project all of its credibility and undisputed success [Press notice].

The company Bettergy, founded in 2008[7], is seeking to develop a low-cost battery for electric vehicles using a solid, non-flammable electrolyte. It has not published any technical information about its technology.

The company Ilika has been working with Toyota since 2008 on the development of solid-state batteries. Specialised in micro-batteries, it has created a high-throughput physical vapor deposition (HT-PVD) platform that has allowed Toyota to develop electrodes and electrolyte materials for lithium-ion batteries much quicker and more efficiently than with traditional materials' discovery processes. Ilika's HT-PVD platform, in combination with other techniques, has enabled the rapid fabrication of a large number of different, structurally related materials in a few automated steps. All of the elements are deposited simultaneously and the composition profile can be carefully carried across the substrate in a controlled manner.

One of the available examples is the synthesis of $Li_{3x}La_{2/3-x}TiO_3$. The choice of substrate proved to be important ($Si/SiO_2/TiO_2/Pt$ (SSTOP), Si/Pt, and $Si/SiO_2/TiO_2/Pt/SrRuO_3$ (SSTOP/SRO)), as well as the lithium content and the annealing temperature of the thin films obtained.

Ilika's technology, which has enabled the successful commercialization of micro-batteries, can be scaled up to produce larger batteries. It could thus lead to smartphone batteries with longer lifetimes, half the size of current batteries, and which could be fully recharged in about 10 min. Ilika

7 Retrieved September 25, 2015, from http://www.bettergy.com/technology.asp

announced that this scale-up could also entail larger batteries for automobile energy storage applications.

Founded in 2007, the company Seeo (http://seeo.com) has focused its research on a solid polymer electrolyte they named DryLyte. DryLyte was initially developed at the Lawrence Berkeley National Laboratory with financial support from the American federal government.

By incorporating high energy electrodes and the Drylite solid electrolyte, DryLite batteries exhibit greater specific energies and extended lifetimes than conventional Li-ion batteries (Figure 6.9)[8].

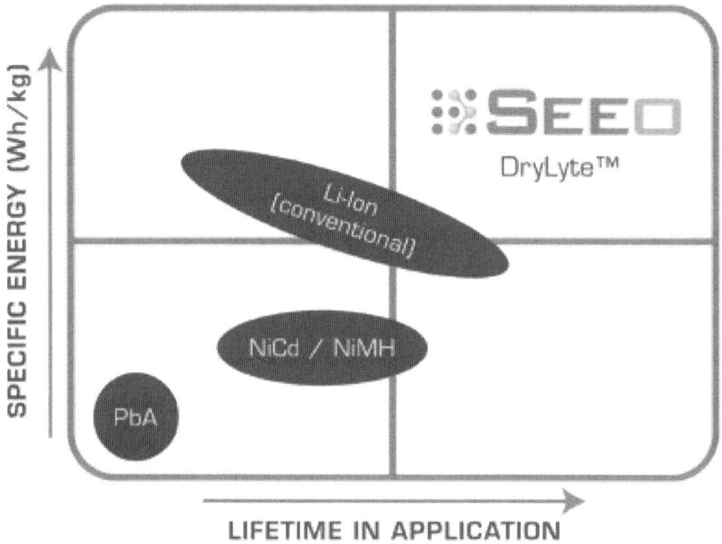

Figure 6.9. *Specific energy and lifetime in application for the DryLite technology compared to other existing technologies [SEE 15]*

SEEO produces cells with an energy density of 220 Wh/kg as well as modules composed of multiple cells for a variety of applications. Standard modules reach an energy density of around 130 Wh/kg (Figure 6.10)[9].

8 Retrieved September 25, 2015, from http://www.seeo.com/drylyte-battery-advntgs/
9 Retrieved September 27, 2015, from http://www.energy.gov/sites/prod/files/2014/07/f17/es129_eitouni_2014_p.pdf

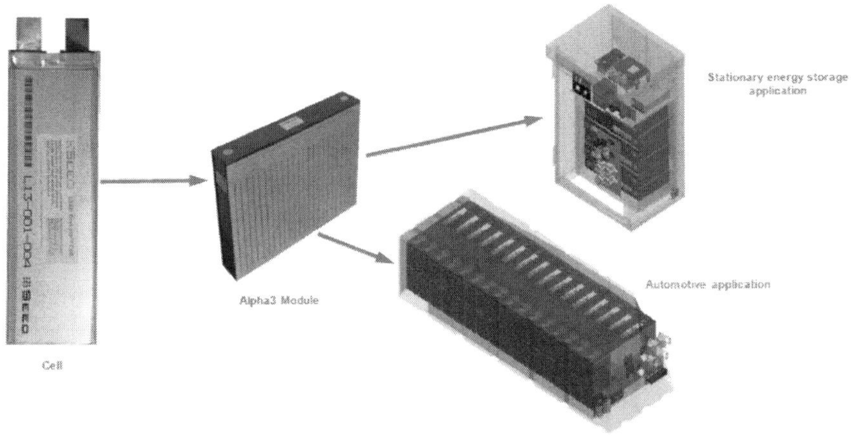

Figure 6.10. *Cells and modules developed by SEEO for stationary and transport-focused applications [EIT 14]*

An outline schematic of a cell using the DryLyte solid polymer electrolyte is shown in Figure 6.11.

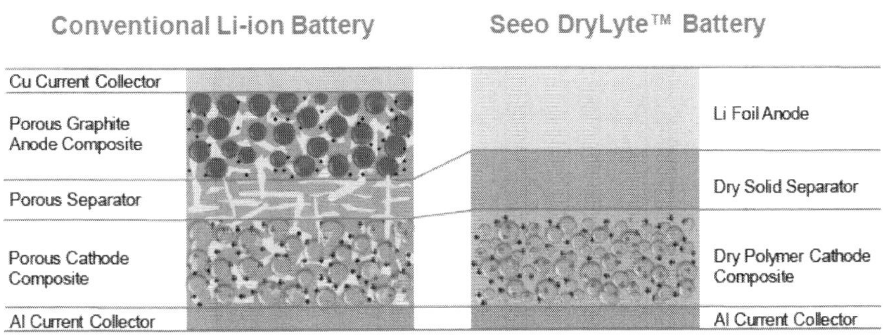

Figure 6.11. *Comparison between a conventional Li-ion battery and a battery using the DryLyte solid polymer electrolyte [EIT 14]. For a color version of this figure, see www.iste.co.uk/viallet/batteries.zip*

The active material used in the composite positive electrode is NCA (Figure 6.12).

• Coated NCA, 5µm zoom

Figure 6.12. *Microscope image of $LiNi_{0.8}Co_{0.15}Al_{0.05}O_2$ (NCA) particles, protected by a coating, used as the active material in the composite positive electrode [EIT 14]*

SEEO announced that they had achieved an energy density of 350 Wh/kg. Bosch has invested several million dollars in the company and counts on benefitting from its technology. Having been almost absent from the market until now, following the failure of its joint-venture with Samsung, the purchase of SEEO clearly suggests the manufacturer's return to the battery world. In fact, Bosch's objective is to be able to offer a battery pack with the same weight as current batteries but with an energy capacity between 50 and 100% greater.

Signed in mid-August 2015 for an undisclosed sum, this purchase importantly allows Bosch to recover SEEO's licences and patents and to target production of a new generation of batteries on an industrial scale [10,11].

The announcement made by Bosch on September 24, 2015[12] regarding this new solid-state battery technology for electric cars indicated that this battery could be ready for production 5 years from now and would offer

10 Retrieved August 27, 2015, from http://www.ibtimes.co.uk/bosch-buys-advanced-battery-start-seeo-targeting-high-growth-electric-car-market-1517245

11 Retrieved September 24, 2015, from http://www.ibtimes.co.uk/bosch-buys-advanced-battery-start-seeo-targeting-high-growth-electric-car-market-1517245 (September 1, 2015 | by Michaël TORREGROSSA)

12 Retrieved October 2, 2015, from http://www.tyrepress.com/2015/09/double-the-range-and-lower-cost-groundbreaking-ev-battery-technology-comes-to-bosch/

more than double the energy density of current products at a lower cost. Bosch confirmed that this advance was made possible by the recent acquisition of the start-up Seeo Inc. and that it now has the first cells needed to continue its development.

The company QuantumScape (http://www.quantumscape.com/) is a start-up spun out from Stanford University. According to their patents, QuantumScape is working on two types of technology. The first is based upon anti-perovskite type solid electrolytes. The second is based upon an AEB battery (for All Electron Battery) in which the charge transport in the electrolyte is mediated by electrons rather than lithium ions.

In March 2015, the Volkswagen group acquired 5% of the shares in QuantumScape, with the option to acquire more, with the aim of integrating QuantumScape's technology into their future generation of electric vehicles[13].

The R&D company Sakti3 (http://sakti3.com/), founded in 2007, originated from MIT and the University of Michigan. Sakti means "power" in Sanskrit and 3 is the atomic number for lithium in the periodic table of elements. The research team led by Ann-Marie Sastry started by designing simulation software able to calculate the ideal combination of the best materials to use to obtain solid-state batteries optimized for performance and cost.

They then set out to define the optimal fabrication process, in terms of ease, speed and cost. The company can fabricate solid-state batteries using a printing technology similar to lithography and processes borrowed from solar cell fabrication.

They have succeeded in demonstrating an energy density of 1162 Wh/L for batteries in the laboratory. Sakti3 has attracted plenty of interest: groups like General Motors, Itochu (a Japanese industrial heavyweight) and Dyson, as well as venture capital firms, have invested tens of millions of dollars, the latest being Dyson with at least $15 million.

The R&D company/laboratory Pathion (http://www.pathion.com/) is active in the research and development of lithium-sulfur and sodium-ion

13 Retrieved September 25, 2015, from http://www.greencarcongress.com/2014/12/20141209-qs.html

batteries. It is more of a scientific laboratory than a company. They are also working on anti-perovskite type solid electrolytes[14].

The company Ceramatec is currently developing solid electrolyte batteries using ceramic membrane separators[15]. Ceramatec is a sub-contractor for a lithium battery program financed by the government and develops sodium and lithium batteries in collaboration with a strategic partner. In August 2013, Ceramatec was selected by ARPA-E's GAMME program to develop a non-porous ceramic membrane with high conductivity for lithium-sulfur batteries in order to reduce self-discharge as much as possible.

Toyota has sold by far the most hybrid vehicles due to the success of its Prius model and is considered by the industry to be the leader in this sector.

In November 2010, Toyota unveiled a prototype solid-state battery, a laminated 10 × 10 cm^2 cell with a potential of 16.26 V (Figure 6.13)[16]. Each cell comprises 4 units of 3.6 V, for a total of 14.4 V. Each unit is a stack consisting of a positive electrode ($LiCoO_2$), a solid electrolyte (Li_2S-P_2S_5) and a negative graphite electrode. Toyota confirmed that the prototype can work at 100 °C.

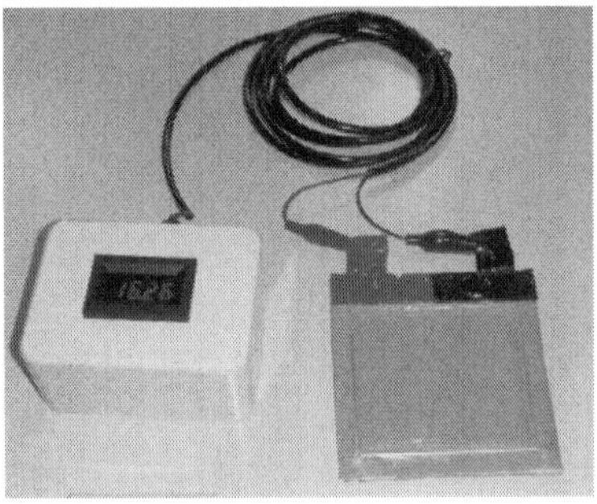

Figure 6.13. *Solid-state cell developed by Toyota and unveiled in November 2010, composed of 4 units of 3.6 V. Each unit is a stack consisting of a positive electrode ($LiCoO_2$), a solid electrolyte (Li_2S–P_2S_5) and a negative graphite electrode*

14 Retrieved September 25, 2015, from http://www.greencarcongress.com/2015/04/20150428-pathion.html

15 Retrieved September 25, 2015, from http://www.ceramatec.com/applications/energy-storage.php

16 Retrieved September 25, 2015, from http://techon.nikkeibp.co.jp/english/NEWS_EN/20101122/187553/

In September 2012, Toyota presented a prototype exhibiting a density five times greater than its predecessor [SHI 12]. This improvement was based upon the use of a $Li_{10}GeP_2S_{12}$ electrolyte, with a conductivity of 1×10^{-2} S/cm. Each cell is a stack of seven positive electrode/electrolyte/negative electrode tri-layers; each of these tri-layers outputs a voltage of around 4V. The voltage of the cell is hence around 28V. The dimensions of the electrodes and solid electrolyte are approximately $70 \times 105 \times 2.5–3$ mm^3. The negative electrode is still graphite but the positive electrode is $Li(Ni_{1/3}Mn_{1/3}Co_{1/3})O_2$. A total of 18 batteries have been connected in parallel and integrated into an electric scooter (Figure 6.14).

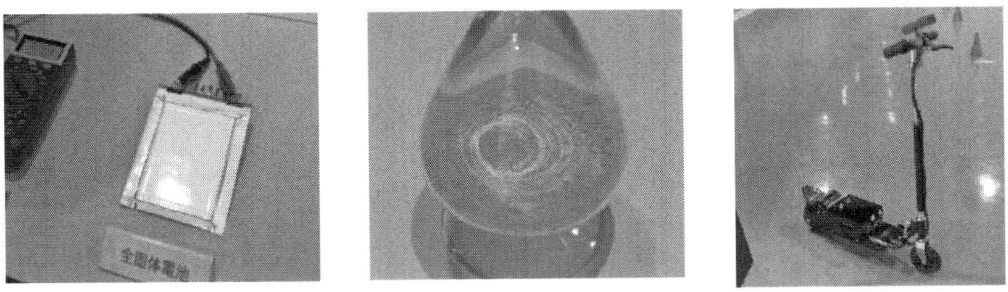

(a) (b) (c)

Figure 6.14. *Prototype of the "solid-state" battery presented by Toyota with a voltage of 28V (from an assembly of seven cells) (a), $Li_{10}GeP_2S_{12}$ electrolyte (b), and a system of 18 batteries connected in parallel and incorporated into an electric scooter (c)*

6.3 Conclusion

In conclusion, solid-state batteries appear to be close to mass commercialization. In any case, there are more and more announcements in this area.

One of the significant challenges for next generation battery companies is to find investors willing to be patient and take the traditional technical risk in companies selling materials. Developing new materials and fabricating them at scale can take many years and, if a new production facility is needed, can require lots of capital.

Conclusion

Growing energy demands require storage systems with ever-improving performance, whether for stationary or automotive and portable power devices. Lithium ion accumulators remain the most effective electrochemical rechargeable batteries and can be split into three categories: classical batteries (with liquid electrolyte), polymer batteries (with a polymer electrolyte of viscosity dependent upon the temperature, which occasionally leads to a disingenuous qualification as solid-state batteries) and solid-state batteries (with solid electrodes and electrolyte).

Here, we are interested in large solid-state batteries with electrolyte layer thicknesses in the order of hundreds of micrometers and not micro-batteries in which the same layer is in the order of hundreds of nanometers to several microns.

Solid-state batteries offer properties that can overcome some of the limitations intrinsic to classical Li-ion batteries with liquid electrolytes:

– first of all, they resolve the safety issues inherent to liquid electrolytes, which can decompose on contact with air, humidity or at temperatures above 60 °C;

– solid electrolytes raise the possibility of high-temperature applications, such as on oil drilling sites. These would be impossible with organic liquid electrolytes due to their boiling points that fall between 90 and 250 °C;

– solid electrolytes should allow lithium metal to be used for the negative electrode;

– the use of solid electrolytes could simplify the processing and assembly of cells since the presence of liquid electrolyte, even if simply impregnated in a separator, places constraints on these operations.

Numerous research groups have started to work on the so-called monolithic solid-state batteries, composed of layers several hundreds of microns in thickness. This thickness must not become too large, as this can lead to a significant increase in internal resistances and hence greater polarization, leading in time to a decrease in performance.

The use of a solid electrolyte in an electrochemical accumulator is not a new idea, with the first studies dating back to the beginning of the 20th Century with the work of Tubandt and Lorenz on the silver conductor α-AgI. Solid electrolytes regained interest in the 1960s due to the discovery of β-aluminum oxide, a sodium ion conductor, which enabled the development of the first "solid-state" batteries based on molten sodium: the Na–S and ZEBRA (for Zero Emissions Batteries Research Activity) batteries. These batteries, exhibiting energy densities between 140 and 240 Wh/kg, use cheap, abundant materials, and close to 90% of the total mass of the battery is recyclable. Their main problem is that they must be used at temperatures between 270 and 350 °C, and hence require an embedded heating system, significantly decreasing their energy density.

Despite active research into this solid-state technology, it has still not been commercialized and two major technological obstacles, common to all types of solid-state design, are holding back the development of such batteries: the weak intrinsic ionic conductivity of solids and the difficulty in obtaining good solid/solid interfaces.

Ionic conductivity, in the electrode materials as well as in the electrolyte, is a key factor in the operation of a battery. It relates to the mobility of the ions in the whole battery and heavily influences the performance, especially the battery power. This conductivity depends on the conductivities in the three layers of the battery, themselves dependent on the intrinsic material conductivities and the conductivities at the grain boundaries, which can in turn be related to the solid/solid interfaces. However, in general, the intrinsic conductivity of solid lithium or sodium conductors at ambient temperature remains much lower than for liquid electrolytes, in the region of 10^{-2} S/cm. Overcoming this problem is hence completely dependent on advances in the research of ionic conductors. The choice of solid electrolyte is critical and is made according to chemical and thermal stability, ion transport number and mechanical resistance. The solid electrolyte must exhibit as high an ionic conductivity as possible, as well as a wide window of electrochemical stability. Li- and Na-ion conductors have thus been studied in various fields of research, ranging from ceramics to polymers, including glasses and glass-ceramics.

– With regard to oxides, the best ionic conductors belong to the LISICON ($Li_{14}ZnGe_4O_{16}$) family, the lithiated NASICONs ($Li_{1+x}Ti_{2-x}M_x(PO_4)_3$; M= Al, Ge, In, Sc) and the garnets ($Li_5La_3M_2O_{12}$; M= Ta, Nb, Bi, Sb).

– The substitution of the oxide ion, O^{2-}, with the more polarizable sulfur ion, S^{2-}, allowed for the discovery of very good ionic conductors in the Thio-LISICON family: $Li_{4+x}Ge_{1-x}P_xS_4$, as well as a solid electrolyte whose ionic conductivity is greater than a conventional liquid electrolyte's – the compound $Li_{10}GeP_2S_{12}$, exhibiting a conductivity of 12×10^{-3} S/cm at 298 K.

– Alongside the LISICONs, conducting glasses were developed by French groups in the 1980s, followed by Japanese groups. The binaries $Li_2S–P_2S_5$ and $Li_2S–SiS_2$ have been the subject of intensive research since the start of the 1980s. Mechanosynthesis has recently enabled the synthesis of better-conducting forms of already-known systems.

– Argyrodites, with formulae $Li_{7-x}PCh_{6-x}X_x$ (Ch = O, S, Se; X = Cl, Br and I), have been used with success in solid-state devices. However, the use of sulfur means the relevant electrolytes must be handled in a controlled atmosphere, which causes problems.

Managing the interfaces between the solid electrolyte and the active material in the electrode is critical. Two types of interface must be considered, the interface between the composite electrode and the solid electrolyte on one hand, and the interfaces between the solid electrolyte and the active material in a composite electrode on the other. These interfaces are not ideal and their properties depend on the materials selected and the fabrication technique used. Just as for classical batteries, it has been shown that an incompatibility between the electrode material and the electrolyte material can lead to the formation of a relatively stable and resistive interface (SEI). Furthermore, even without reactivity, poor grain interfaces (whether between the same or different materials) increase the grain boundary resistance, which worsens the overall conductivity of the battery. This is a complex problem to resolve, especially since it depends on the materials and on the fabrication technique used, which can in turn impose their own constraints.

– In order to obtain good solid/solid interfaces, it is imperative to match the fabrication technique to the characteristics of the chosen materials.

– Sulfur-based electrolytes are unusual in that they are easily deformed and offer the potential for fabrication of compact stacks by simple work hardening.

– Ceramic oxide electrolytes, which are harder, require high temperature densification such as assembly by spark plasma sintering (SPS), hot pressing (HP) or pulsed laser deposition (PLD).

Solid-state batteries were developed from the start of the 1980s, but it was the work by the Takada and Tatsumisago groups in the 1990s that led to the emergence of a new generation of solid-state batteries, using $LiCoO_2$ as the positive electrode. The discovery of the glass-ceramic $80Li_2S–20P_2S_5$ raised the possibility of a variety of materials for the negative electrode:

– $Cu_xMo_6S_{8-y}$, for example, distinguishes itself with a stable capacity of 100 mAh/g over 2000 cycles;

– metallic lithium is the ideal candidate but its use, even in a solid-state device, is limited to low current densities – especially for $LiCoO_2$ –– and indium or Li-In alloys give better results;

– the compound $Li_4Ti_5O_{12}$ seems to be a preferred material for the negative electrode in solid-state batteries. Without any volume expansion during the insertion/deinsertion phases, it can withstand extremely high cycling regimes.

One promising technique for applying the spark plasma sintering (SPS) process to the fabrication of solid-state batteries is still very recent, the first publication by Aboulaich *et al.* dating from 2011. Use of the SPS technique allowed the fabrication of very compact solid-state batteries with very good interfaces, including in composite electrodes that can reach thicknesses of more than 500μm.

Areas for future advances include the following:

– coating the active material with an electrolyte layer; in some cases, coatings have enabled the fabrication of an electrode containing 90% by mass of the active material;

– optimization of the pressure applied when the battery is in use, rarely considered but essential for good performance.

A large number of challenges therefore persist at the scientific and technological level. The development of electrolytes exhibiting good ionic conductivity combined with chemical and electrochemical stability and controlled thicknesses, as well as cells with thick electrodes and optimized interfaces between the electrodes and electrolyte, continue to be the subject of intensive research. Research also continues into developing practical solutions for the industry. Few companies have commercialized solid electrolytes, but there has recently been an increase in publicity announcing the commercialization of solid-state batteries.

The research effort in this field is intensifying and the number of publications is growing. A range of review articles covering solid electrolytes or solid-state batteries have been published in the last 3 years, which are likely to be useful resources to complement the information provided in this book:

– Recent Advances in Inorganic Solid Electrolytes for Lithium Batteries [CAO 14];

– All-solid-state Lithium Batteries with Inorganic Solid Electrolytes: Review of Fundamental Science [YAO 16];

– Lithium Battery Chemistries enabled by Solid-state Electrolytes [MAN 17];

– Relevance of Solid Electrolytes for Lithium-based Batteries: A Realistic View [LOT 17];

– Practical Challenges Hindering the Development of Solid State Li Ion Batteries [KER 17];

– recent advances in all-solid-state rechargeable lithium batteries [SUN 17].

Bibliography

[ABO 11] ABOULAICH A., BOUCHET R., DELAIZIR G. *et al.*, "A new approach to develop safe allinorganic monolithic Li-ion batteries", *Advanced Energy Materials*, vol. 1, no. 2, pp. 179–183, 2011.

[AGR 04] AGRAWAL R.C., VERMA M.L., GUPTA R.K., "Electrical and electrochemical properties of a new silver tungstate glass system: x[0.75AgI:0.25AgCl]: $(1 - x)$ [Ag_2O:WO_3]", *Solid State Ionics*, vol. 171, pp. 199–204, 2004.

[AGR 08] AGRAWAL R.C., PANDEY G.P., "Solid polymer electrolytes: materials designing and all-solid-state battery applications: an overview", *Journal of Physics D: Applied Physics*, vol. 41, no. 22, p. 223001, 2008.

[AHN 89] AHN B.T., HUGGINS R.A., "Synthesis and lithium conductivities of Li_2SiS_3 and Li_4SiS_4", *Materials Research Bulletin*, vol. 24, no. 7, pp. 889–897, 1989.

[AHN 05] AHN J.K., YOON S.G., "Characteristics of amorphous lithium lanthanum titanate electrolyte thin films grown by PLD for use in rechargeable lithium microbatteries", *Electrochemical and Solid-State Letters*, vol. 8, no. 2, pp. A75–A78, 2005.

[AIS 10] AIST, Fabrication of an All-solid-state Thin-film Lithium-ion Battery Prototype Using a Room Temperature Process, available at: http://www.aist.go.jp/aist_e/list/latest_research/2010/20101224/20101224.html, consulted on 08/27/2013, 24 December, 2010.

[AKR 86] AKRIDGE J.R., VOURLIS H., "Solid state batteries using vitreous solid electrolytes", *Solid State Ionics*, vol. 18, pp. 1082–1087, 1986.

[AKR 88] AKRIDGE J.R., VOURLIS H., "Performance of Li/TiS_2 solid state batteries using phosphorous chalcogenide network former glasses as solid electrolyte", *Solid State Ionics*, vol. 28, pp. 841–846, 1988.

[ALP 79] ALPEN U.V., BELL M.F., WICHELHAUS W., "Phase-transition in nasicon ($Na_3Zr_2Si_2Po_{12}$)", *Materials Research Bulletin*, vol. 14, no. 10, pp. 1317–1322, 1979.

[AMA 96] AMATUCCI G.G., TARASCON J.M., KLEIN L.C., "Cobalt dissolution in $LiCoO_2$-based non-aqueous rechargeable batteries", *Solid State Ionics*, vol. 83, nos 1–2, pp. 167–173, 1996.

[AON 89] AONO H., SUGIMOTO E., SADAAKA Y. *et al.*, "Ionic conductivity of the lithium titanium phosphate ($Li_{1+x}M_xTi_{2-x}(PO_4)_3$, M= Al, Sc, Y, and La) systems", *Journal of the Electrochemical Society*, vol. 136, no. 2, pp. 590–591, 1989.

[AON 90] AONO H., SUGIMOTO E., SADAOKA Y. *et al.* "Ionic conductivity of solid electrolytes based on lithium titanium phosphate", *Journal of the electrochemical society*, vol. 137, no. 4, pp. 1023–1027, 1990.

[AOT 94] AOTANI N., IWAMOTO K., TAKADA K. *et al.*, "Synthesis and electrochemical properties of lithium ion conductive glass, Li_3PO_4–Li_2S–SiS_2", *Solid State Ionics*, vol. 68, nos 1–2, pp. 35–39, 1994.

[APP 00a] APPETECCHI G.B., PASSERINI S., "PEO-carbon composite lithium polymer electrolyte", *Electrochimica Acta*, vol. 45, pp. 2139–2145, 2000.

[APP 00b] APPETECCHI G.B., CROCE F., PERSI L. *et al.*, "Transport and interfacial properties of composite polymer electrolytes", *Electrochimica Acta*, vol. 45, pp. 1481–1490, 2000.

[APP 00c] APPETECCHI G.B., SCACCIA S., PASSERINI S., "Investigation on the stability of the lithium-polymer electrolyte interface", *Journal of the Electrochemical Society*, vol. 147, pp. 4448–4452, 2000.

[APP 01a] APPETECCHI G.B., ALESSANDRINI F., DUAN R.G. *et al.*, "Electrochemical testing of industrially produced PEO-based polymer electrolytes", *Journal of Power Sources*, vol. 101, pp. 42–46, 2001.

[APP 01b] APPETECCHI G.B., HENDERSON W., VILLANO P. *et al.*, "PEO-LiN (SO_2 CF_2 CF3)$_2$ polymer electrolytes: I. XRD, DSC, and ionic conductivity characterization", *Journal of The Electrochemical Society*, vol. 148, pp. A1171–A1178, 2001.

[ARA 97] ARAGANE J., MATSUI K., ANDOH H. *et al.*, "Development of 10 Wh class lithium secondary cells in the 'New Sunshine Program", *Journal of Power Sources*, vol. 68, pp. 13–18, 1997.

[ARB 07] ARBI K., ROJO J.M., SANZ J., "Lithium mobility in titanium based Nasicon $Li_{1+x}Ti_{2-x}Al_x(PO_4)_3$ and $LiTi_{2-x}Zr_x(PO_4)_3$ materials followed by NMR and impedance spectroscopy", *Journal of the European Ceramic Society*, vol. 27, no. 13, pp. 4215–4218, 2007.

[ARM 78] ARMAND M.B, CHABAGNO J.M., DUCLOT M.J., *Extended Abstracts of Second International Meeting on Solid Electrolytes*, St. Andrews, Scotland, 20–22 September 1978.

[ARM 79] ARMAND M.B., CHABAGNO J.M., DUCLOT M.J., in VASHISHTA P. (ed.), *Fast Transport in Solids*, North Holland, NY, p. 131, 1979.

[ARM 89] ARMAND M., CHABAGNO J.M., DUCLOT M.J. *et al.* (eds), *Fast Ion Transport in Solids*, Elsevier, New York, 1989.

[ASO 10] ASO K., KITAURA H., HAYASHI A. *et al.*, "$SnP_{0.94}$ active material synthesized in high-boiling solvents for all-solid-state lithium batteries", *Journal of the Ceramic Society of Japan*, vol. 118, no. 1379, pp. 620–622, 2010.

[AVE 10] AVEM, "Premier site d'information sur le véhicule électrique et hybride", available at: http://www.avem.fr/?page=batterie&cat=technos&kind=zebra, consulted on 29/08/2017, 2010.

[AZU 99] AZUMA H., IMOTO H., YAMADA S.I. *et al.*, "Advanced carbon anode materials for lithium ion cells", *Journal of Power Sources*, vol. 81, pp. 1–7, 1999.

[BAB 99] BABA M., KUMAGAI N., KOBAYASHI H. *et al.*, "Fabrication and electrochemical characteristics of all-solid-state lithium-ion batteries using V_2O_5 thin films for both electrodes", *Electrochemical and Solid-State Letters*, vol. 2, no. 7, pp. 320–322, 1999.

[BAT 92] BATES J.B., DUDNEY N.J., GRUZALSKI G.R. *et al.*, "Electrical properties of amorphous lithium electrolyte thin films", *Solid State Ionics*, vol. 53, pp. 647–654, 1992.

[BAT 95] BATES J. B., LUBBEN D., DUDNEY N. J., "Thin-film Li-$LiMn_2O_4$ batteries", *IEEE Aerospace and Electronic Systems Magazine*, vol. 10, no. 4, pp. 30–32, 1995.

[BEA 95] BEATON S.P., BISHOP G.A., ZIANG Y. *et al.*, "On-Road Vehicle Emissions: Regulations, Costs, and Benefits", *Science*, vol. 268, no. 5213, pp. 991–993, 1995.

[BEE 37] BEEVERS C.A., ROSS M.A.S., "The crystal structure of "beta alumina" $Na_2O{\cdot}11Al_2O_3$", in *Zeitschrift für Kristallographie – Crystalline Materials*, vol. 97, pp. 59–66, 1937.

[BFM 15] BFM Business, "Renault: l'autonomie de la Zoé va passer à 300 km", available at : http://bfmbusiness.bfmtv.com/entreprise/renault-l-autonomie-de-la-zoe-va-passer-a-300-km-859555.html, consulted on 25/09/2015, 2015.

[BOE 13] BOEKE H.E., *Zeitschrift Fur Krystallographie Und Mineralogie*, vol. 53, no. 2, p. 149, 1913.

[BOR 95] BORGHINI M.C., MASTRAGOSTINO M., PASSERINI S. *et al.*, "Electrochemical Properties of Polyethylene oxide–Li [(CF3 SO2)2 N] –Gamma-LiAlO2 composite polymer electrolytes", *Journal of the Electrochemical Society*, vol. 142, pp. 2118–2121, 1995.

[BOU 12] BOULINEAU S., COURTY M., TARASCON J.-M. *et al.*, "Mechanochemical synthesis of Li-argyrodite Li_6PS_5X (X = Cl, Br, I) as sulfur-based solid electrolytes for all solid state batteries application", *Solid State Ionics*, vol. 221, pp. 1–5, 2012.

[BOU 13a] BOULINEAU S., Synthèse et caractérisation de matériaux céramiques, vitreux et vitrocéramique à base de soufre, utilisables comme électrolytes dans les batteries 'tout solide', PhD Thesis, Université de Picardie Jules Verne, France, 2013.

[BOU 13b] BOULINEAU S., TARASCON J.-M., LERICHE J.-B. *et al.*, "Electrochemical properties of all-solid-state lithium secondary batteries using Li-argyrodite Li_6PS_5Cl as solid electrolyte", *Solid State Ionics*, vol. 242, pp. 45–48, 2013.

[BRA 12] BRAUN P.V., CHO J., PIKUL J.H. *et al.* , "High power rechargeable batteries", *Current Opinion in Solid State and Materials Science*, vol. 16, no. 4, pp. 186–198, 2012.

[BRA 13] BRAGA M.H., STOCKHAUSEN V., OLIVEIRA J. *et al.*,"The role of defects in Li_3ClO solid electrolyte: calculations and experiments", *MRS Online Proceedings Library Archive, 2012 MRS Fall Meeting*, vol.1526, 2013.

[BRA 14] BRAGA M.H., FERREIRA J.A., STOCKHAUSEN V. *et al.*, "Novel Li_3ClO based glasses with superionic properties for lithium batteries", *Journal of Materials Chemistry A*, vol. 2, no. 15, pp. 5470–5480, 2014.

[BRA 16] BRAGA M.H., MURCHISON A.J., FERREIRA J.A. *et al.*, "Glass-amorphous alkali-ion solid electrolytes and their performance in symmetrical cells", *Energy & Environmental Science*, vol. 9, no. 3, pp. 948–954, 2016.

[BRU 80] BRUCE P.G., WEST A.R., "Phase diagram of the LISICON, solid electrolyte system, Li_4GeO_4-Zn_2GeO_4", *Materials Research Bulletin*, vol. 15, no. 3, pp. 379–385, 1980.

[BRU 84] BRUCE P.G., WEST A.R., "Ion trapping and its effect on the conductivity of LISICON and other solid electrolytes", *Journal of Solid State Chemistry*, vol. 53, no. 3, pp. 430–434, 1984.

[BRU 92] BRUCE P.G., HARDGRAVE M.T., VINCENT C.A., "The determination of transference numbers in solid polymer electrolytes using the Hittorf method", *Solid State Ionics*, vol. 53, pp. 1087–1094, 1992.

[BRU 97] BRUCE P.G., "Solid-state chemistry of lithium power sources", *Chemical Communications*, vol. 19, pp. 1817–1824, 1997.

[BRU 12] BRUCE P.G., FREUNBERGER S.A., HARDWICK I.J. *et al.*, "Li-O2 and Li-S batteries with high energy storage", *Nature Materials*, vol. 11, pp. 19–29, 2012.

[BUL 02] BULLOCK S.E., KOFINAS P., "Polymeric nanoscale all-solid state battery", *Materials Research Society Symposium Proceedings*, vol. 705, pp. 223–226, 2002.

[CAO 14] CAO C., LI Z.B., WANG X.L. *et al.*, "Recent advances in inorganic solid electrolytes for lithium batteries", *Frontiers in Energy Research*, vol. 2, no. 25, pp. 1–10, 2014.

[CAP 91] CAPUANO F., CROCE F., SCROSATI B., "Composite polymer electrolytes", *Journal of the Electrochemical Society*, vol. 138, pp. 1918–1922, 1991.

[CAR 83] CARETTE B., MAURIN M., RIBES M. *et al.*, "Ionic conductive sulfide-based M2S–GeS2–MI (M = Li, Ag) glass systems. Their use in solid state batteries", *Solid State Ionics*, vols 9–10, p. 655, 1983.

[CAS 17] CASSEL A., FLEUTOT B., COURTY M. *et al.*, "Sol-gel synthesis and electrochemical properties extracted by phase inflection detection method of NASICON-type solid electrolytes LiZr2(PO4)3 and Li1.2Zr1.9Ca0.1(PO4)3", *Solid State Ionics*, vol. 309, pp. 63–70, 2017.

[CED 11] CEDER G., "Chimie théorique combinatoire pour la quête de nouveaux matériaux", Collège de France, seminar, January 26, 2011, https://annuaire-cdf.revues.org/1629.

[CHE 13] CHEN J., "Recent progress in advanced materials for lithium ion batteries", *Materials*, vol. 6, no. 1, pp. 156–183, 2013.

[CHI 12] CHIKU M., TSUJIWAKI W., HIGUCHI E. *et al.*, "Microelectrode studies on kinetics of charge transfer at an interface of Li metal and Li2S–P2S5 solid electrolytes", *Electrochemistry*, vol. 80, no. 10, pp. 740–742, 2012.

[CHO 13] CHO Y.J., IM H.S., MYUNG Y. *et al.*, "Germanium sulfide (II and IV) nanoparticles for enhanced performance of lithium ion batteries", *Chemical Communications*, vol. 49, no. 41, pp. 4661–4663, 2013.

[CRE 89] CREUS R., SARRADIN J., ASTIER R. *et al.*, "The use of ionic and mixed conductive glasses in microbatteries", *Materials Science and Engineering: B*, vol. 3, nos 1–2, pp. 109–112, 1989.

[CRE 99] CRETIN M., FABRY P., "Comparative study of lithium ion conductors in the system Li1+ xAlxA2−xIV(PO4)3 with AIV= Ti or Ge and $0 \leq x \leq 0{\cdot}7$ for use as Li+ sensitive membranes", *Journal of the European Ceramic Society*, vol. 19, no. 16, pp. 2931–2940, 1999.

[CRO 98] CROCE F., APPETECCHI G.B., PERSI L. *et al.*, "Nanocomposite polymer electrolytes for lithium batteries", *Nature*, vol. 394, pp. 456–458, 1998.

[CUS 06] CUSSEN E.J., "The structure of lithium garnets: cation disorder and clustering in a new family of fast Li^+ conductors", *Chemical Communications*, vol. 4, pp. 412–413, 2006.

[DAS 03] DAS S.S., BARANWAL B.P., GUPTA C.P. *et al.*, "Characteristics of solid-state batteries with zinc/cadmium halide-doped silver phosphate glasses as electrolytes", *Journal of Power Sources*, vol. 114, no. 2, pp. 346–351, 2003.

[DAS 16] DAS S., NGENE P., NORBY P. *et al.*, "All-solid-state lithium-sulfur battery based on a nanoconfined $LiBH_4$ electrolyte", *Journal of the Electrochemical Society*, vol. 163, pp. A2029–A2034, 2016.

[DEI 08] DEISEROTH H.J., KONG S.T., ECKERT H. *et al.*, "Li_6PS_5X: a class of crystalline Li-rich solids with an unusually high Li^+ mobility", *Angewandte Chemie International Edition*, vol. 47, no. 4, pp. 755–758, 2008.

[DEL 88] DELMAS C., NADIRI A., SOUBEYROUX J.L., "The nasicon-type titanium phosphates $ATi_2(PO_4)_3$ (A= Li, Na) as electrode materials", *Solid State Ionics*, vol. 28, pp. 419–423, 1988.

[DEL 95] DELMAS C., COGNAC-ARADOU H., "Formation of the ω-type phase by lithium intercalation in (Mo, V) oxides deriving from V_2O_5", *Journal of Power Sources*, vol. 54, p. 406, 1995.

[DEL 12a] DELAIZIR G., MANAFI N., JOUAN G. *et al.*, "All-solid-state silver batteries assembled by Spark Plasma Sintering", *Solid State Ionics*, vol. 207, pp. 57–63, 2012.

[DEL 12b] DELAIZIR G., VIALLET V., ABOULAICH A. *et al.*, "Batteries: the stone age revisited: building a monolithic inorganic lithium-ion battery", *Advanced Functional Materials*, vol. 22, no. 10, pp. 2140–2147, 2012.

[DIA 00] DIAS F.B., PLOMP L., VELDHUIS J.B., "Trends in polymer electrolytes for secondary lithium batteries", *Journal of Power Sources*, vol. 88, no. 2, pp. 169–191, 2000.

[DOL 02] DOLLE M., Etude par spectroscopie d'impédance électrochimique, couplée à la microscopie électronique, d'interfaces de batteries au lithium et à ions lithium, PhD Thesis, Université de Picardie Jules Verne, France, 2002.

[DUD 05] DUDNEY N.J., "Solid-state thin-film rechargeable batteries", *Materials Science and Engineering: B*, vol. 116, no. 3, pp. 245–249, 2005.

[EIT 14] EITOUNI H., "High-Voltage Solid Polymer Batteries for Electric Drive Vehicles", SEEO, available at: https://www.energy.gov/sites/prod/files/2014/07/f17/es129_eitouni_2014_p.pdf, consulted on 27/09/2015, 2014.

[END 00] ENDO M., KIM C., NISHIMURA K. *et al.*, "Recent development of carbon materials for Li ion batteries", *Carbon*, vol. 38, pp. 183–197, 2000.

[EVA 87] EVANS J., VINCENT C.A., BRUCE P.G., "Electrochemical measurement of transference numbers in polymer electrolytes", *Polymer*, vol. 28, no. 13, pp. 2324–2328, 1987.

[FER 10] FERGUS J.W., "Recent developments in cathode materials for lithium ion batteries", *Journal of Power Sources*, vol. 195, no. 4, pp. 939–954, 2010.

[FLE 11a] FLEUTOT B., PECQUENARD B., LE CRAS F. *et al.*, "Characterization of all-solid-state Li/LiPONB/TiOS microbatteries produced at the pilot scale", *Journal of Power Sources*, vol. 196, pp. 10289–10296, 2011.

[FLE 11b] FLEUTOT B., PECQUENARD B., MARTINEZ H. *et al.*, "Investigation of the local structure of LiPON thin films to better understand the role of nitrogen on their performance", *Solid State Ionics*, vol. 186, pp. 29–36, 2011.

[FLE 12] FLEUTOT B., PECQUENARD B., MARTINEZ H., "Thorough study of the local structure of LiPON thin films to better understand the influence of a solder-reflow type thermal treatment on their performances", *Solid State Ionics*, vol. 206, pp. 72–77, 2012.

[FLE 13] FLEUTOT B., PECQUENARD B., MARTINEZ H. *et al.*, "Lithium borophosphate thin film electrolyte as an alternative to LiPON for solder-reflow processed lithium-ion microbatteries", *Solid State Ionics*, vols 249–250, pp. 49–55, 2013.

[GAM 71] GAMBLE F.R., OSIECKI J.H., CAIS M. *et al.*, "Intercalation complexes of Lewis bases and layered sulfides: a large class of new superconductors", *Science*, vol. 174, no. 4008, pp. 493–497, 1971.

[GEI 11] GEIGER C.A., ALEKSEEV E., LAZIC B. *et al.*, "Crystal chemistry and stability of $Li_7La_3Zr_2O_{12}$ garnet: a fast lithium-ion conductor", *Inorganic Chemistry*, vol. 50, no. 3, pp. 1089–1097, 2011.

[GEN 09] GENG H., MEI A., LIN Y. *et al.*, "Effect of sintering atmosphere on ionic conduction and structure of $Li_{0.5}La_{0.5}TiO_3$ solid electrolytes", *Materials Science and Engineering: B*, vol. 164, no. 2, pp. 91–95, 2009.

[GOO 76] GOODENOUGH J.B., HONG H.P., KAFALAS J.A., "Fast Na^+-ion transport in skeleton structures", *Materials Research Bulletin*, vol. 11, no. 2, pp. 203–220, 1976.

[GRA 91] GRAY F.M., *Solid Polymer Electrolytes: Fundamentals and Technological Applications*, Wiley-VCH, Weinheim, 1991.

[GRA 97] GRAY F.M., *Polymer Electrolytes,* Royal Society of Chemistry, Cambridge, 1997.

[GRA 00] GRAY F.M., ARMAND M., OSAKA T. *et al.* (eds), *Energy Storage System for Electronics*, Gordon and Breach, Amsterdam, 2000.

[GUI 15] GUIN M., TIETZ F., "Survey of the transport properties of sodium superionic conductor materials for use in sodium batteries", *Journal of Power Sources*, vol. 273, pp. 1056–1064, 2015.

[HAG 68] HAGMAN L.O., KIERKEGAARD P., KARVONEN P. *et al.*, "The crystal structure of $NaMe_2^{IV}(PO_4)_3$; Me^{IV}= Ge, Ti, Zr", *Acta Chem. Scand*, vol. 22, no. 6, pp. 1822–1832, 1968.

[HAN 15] HAN F., GAO T., ZHU Y. *et al.*, "A battery made from a single material", *Advanced Materials*, vol. 27, no. 23, pp. 3473–3483, 2015.

[HAY 99] HAYASHI A., TATSUMISAGO M., MINAMI T., "Electrochemical properties for the lithium ion conductive (100–x)(0.6Li_2S·0.4SiS_2)·$x$$Li_4SiO_4$ oxysulfide glasses", *Journal of the Electrochemical Society*, vol. 146, no. 9, pp. 3472–3475, 1999.

[HAY 01] HAYASHI A., HAMA S., MORIMOTO H. *et al.*, "Preparation of Li_2S–P_2S_5 amorphous solid electrolytes by mechanical milling", *Journal of the American Ceramic Society*, vol. 84, no. 2, pp. 477–479, 2001.

[HAY 03a] HAYASHI A., HAMA S., MINAMI T. *et al.*, "Formation of superionic crystals from mechanically milled Li_2S–P_2S_5 glasses", *Electrochemistry communications*, vol. 5, no. 2, pp. 111–114, 2003.

[HAY 03b] HAYASHI A., ISHIKAWA Y., HAMA S. *et al.*, "Fast lithium-ion conducting glass-ceramics in the system Li_2S–SiS_2–P_2S_5", *Electrochemical and Solid-State Letters*, vol. 6, no. 3, pp. A47–A49, 2003.

[HAY 03c] HAYASHI A., OHTOMO T., MIZUN, F. *et al.*, "All-solid-state Li/S batteries with highly conductive glass–ceramic electrolytes", *Electrochemistry Communications*, vol. 5, no. 8, pp. 701–705, 2003.

[HAY 05] HAYASHI A., KONISHI T., TADANAGA K. *et al.*, "All-solid-state lithium secondary batteries with SnS–P_2S_5 negative electrodes and Li_2S–P_2S_5 solid electrolytes", *Journal of Power Sources*, vol. 146, no. 1, pp. 496–500, 2005.

[HAY 08a] HAYASHI A., NISHIO Y., KITAURA H. *et al.*, "Novel technique to form electrode–electrolyte nanointerface in all-solid-state rechargeable lithium batteries", *Electrochemistry Communications*, vol. 10, pp. 1860–1863, 2008.

[HAY 08b] HAYASHI A., OHTSUBO R., OHTOMO T. *et al.*, "All-solid-state rechargeable lithium batteries with Li_2S as a positive electrode material", *Journal of Power Sources*, vol. 183, no. 1, pp. 422–426, 2008.

[HAY 09a] HAYASHI A., INOUE A., TATSUMISAGO M., "Electrochemical performance of NiP_2 negative electrodes in all-solid-state lithium secondary batteries", *Journal of Power Sources*, vol. 189, no. 1, pp. 669–671, 2009.

[HAY 09b] HAYASHI A., MINAMI K., TATSUMISAGO M., "High lithium ion conduction of sulfide glass-based solid electrolytes and their application to all-solid-state batteries", *Journal of Non-Crystalline Solids*, vol. 355, no. 37, pp. 1919–1923, 2009.

[HAY 12] HAYASHI A., NOI K., SAKUDA A. *et al.*, "Superionic glass-ceramic electrolytes for room-temperature rechargeable sodium batteries", *Nature Communications*, vol. 3, p. 856, 2012.

[HAY 13] HAYASHI A., TATSUMISAGO M., *International Battery Association (IBA) 2013 Meeting*, Barcelona, Spain, 2013.

[HAY 14] HAYASHI A., NOI K., TANIBATA N. *et al.*, "High sodium ion conductivity of glass–ceramic electrolytes with cubic Na_3PS_4", *Journal of Power Sources*, vol. 258, pp. 420–423, 2014.

[HIR 95] HIRAI K., TATSUMISAGO M., MINAMI T., "Thermal and electrical properties of rapidly quenched glasses in the systems $Li_2S–SiS_2–Li_xMO_y$ (Li_xMO_y=Li_4SiO_4, Li_2SO_4)", *Solid State Ionics*, vol. 78, nos 3–4, pp. 269–273, 1995.

[HON 76a] HONG H.Y.P., "Crystal structures and crystal chemistry in the system $Na_{1+x}Zr_2Si_xP_{3-x}O_{12}$", *Materials Research Bulletin*, vol. 11, pp. 173–182, 1976.

[HON 76b] HONG H.Y.P., GOODENOUGH J.B., KAFALAS J.A., "Fast Na^+-ion transport in skeleton structures", *Materials Research Bulletin*, vol. 11, pp. 203–220, 1976.

[IEO 16] International Energy Outlook 2016, U.S. Energy Information Administration, figure 1.5, p. 9, available at: https://www.eia.gov/outlooks/ieo/pdf/0484(2016).pdf, consulted on 29/08/2017, 2016.

[IGN 05] IGNASZAK A., PASIERB P., GAJERSKI R. *et al.*, "Synthesis and properties of Nasicon-type materials", *Thermochimica Acta*, vol. 426, pp. 7–14, 2005.

[JAN 92] JANSEN M., HENSELER U., "Synthesis, structure determination, and ionic conductivity of sodium tetrathiophosphate", *Journal of Solid State Chemistry*, vol. 99, no. 1, pp. 110–119, 1992.

[JAN 16] JANEK J., ZEIER W.G., "A solid future for battery development", *Nature Energy*, vol. 1, pp. 1–4, 2016.

[JIN 13] JIN Y., MCGINN P.J., "Bulk solid state rechargeable lithium ion battery fabrication with Al-doped $Li_7La_3Zr_2O_{12}$ electrolyte and $Cu_{0.1}V_2O_5$ cathode", *Electrochimica Acta*, vol. 89, pp. 407–412, 2013.

[JOH 58] JOHNSTON W.D., HEIKES R.R., SESTRICH D., "The preparation, crystallography, and magnetic properties of the $Li_xCo_{(1-x)}O$ system", *Journal of Physics and Chemistry of Solids*, vol. 7, no. 1, pp. 1–13, 1958.

[KAL 96] KALHAMMER F.R., KOZAWA A., MOYER C.B. *et al.*, "Electric vehicle batteries", *The Electrochemical Society Interface*, vol. 5, pp. 32–37, 1996.

[KAM 11] KAMAYA N., HOMMA K., YAMAKAWA Y. *et al.*, "A lithium superionic conductor", *Nature Materials*, vol. 10, no. 9, p. 682, 2011.

[KAN 83] KANEHORI K., MATSUMOTO K., MIYAUCHI K. *et al.*, "Thin film solid electrolyte and its application to secondary lithium cell", *Solid State Ionics*, vol. 9, pp. 1445–1448, 1983.

[KAN 01] KANNO R., MURAYAMA M., "Lithium Ionic Conductor Thio-LISICON: The Li2S-GeS2-P2S5 System", *Journal of The Electrochemical Society*, vol. 148, no. 7, pp. A742-A746, 2001.

[KAS 07] KASAVAJJULA U., WANG C., APPLEBY A.J., "Nano-and bulk-silicon-based insertion anodes for lithium-ion secondary cells", *Journal of Power Sources*, vol. 163, no. 2, pp. 1003–1039, 2007.

[KAT 04] KATO Y., YOKOYAMA S., YABE T. *et al.*, "Ionic conductivity and transport number of lithium ion in polymer electrolytes containing PEG–borate ester", *Electrochimica acta*, vol. 50, no. 2, pp. 281–284, 2004.

[KAT 12] KATO Y., KAWAMOTO K., KANNO R. *et al.*, "Discharge performance of all-solid-state battery using a lithium superionic conductor $Li_{10}GeP_2S_{12}$", *Electrochemistry*, vol. 80, no. 10, pp. 749–751, 2012.

[KAT 16] KATO Y, HORI S., SAITO T. *et al.*, "High-power all-solid-state batteries using sulfide superionic conductors", *Nature Energy*, vol. 1, article number: 16030, 2016.

[KEN 87] KENNEDY J.H., YANG Y., "Glass-forming region and structure in SiS–Li_2S–LiX (X= Br, I)", *Journal of Solid State Chemistry*, vol. 69, no. 2, pp. 252–257, 1987.

[KEN 88] KENNEDY J.H., ZHANG Z., "Further characterization of SiS_2–Li_2S glasses doped with lithium halide", *Journal of the Electrochemical Society*, vol. 135, no. 4, pp. 859–862, 1988.

[KER 17] KERMAN K., LUNTZ A., VISWANATHAN V. *et al.*, "Practical challenges hindering the development of solid state li ion batteries", Journal of the Electrochemical Society, vol. 164, no. 7, A1731–A1744, 2017.

[KIM 08] KIM Y., HWANG H., LAWLER K. *et al.*, "Electrochemical behavior of Ge and GeX_2 (X= O, S) glasses: improved reversibility of the reaction of Li with Ge in a sulfide medium", *Electrochimica Acta*, vol. 53, no. 15, pp. 5058–5064, 2008.

[KIM 15] KIM J.G., SON B., MUKHERJEE S. *et al.*, "A review of lithium and non-lithium based solid state batteries", *Journal of Power Sources*, vol. 282, pp. 299–322, 2015.

[KIT 07] KITAURA H., TAKAHASHI K., MIZUNO F. *et al.*, "Preparation of α-Fe_2O_3 electrode materials via solution process and their electrochemical properties in all-solid-state lithium batteries", *Journal of The Electrochemical Society*, vol. 154, no. 7, pp. A725–A729, 2007.

[KIT 09] KITAURA H., HAYASHI A., TADANAGA K. *et al.*, "High-rate performance of all-solid-state lithium secondary batteries using $Li_4Ti_5O_{12}$ electrode", *Journal of Power Sources*, vol. 189, pp. 145–148, 2009.

[KIT 10a] KITAURA H., HAYASHI A., TADANAGA K. *et al.*, "All-solid-state lithium secondary batteries using $LiMn_2O_4$ electrode and Li_2S–P_2S_5 solid electrolyte", *Journal of the Electrochemical Society*, vol. 157, no. 4, pp. A407–A411, 2010.

[KIT 10b] KITAURA H., HAYASHI A., TADANAGA K. *et al.*, "Electrochemical performance of all-solid-state lithium secondary batteries with Li–Ni–Co–Mn oxide positive electrodes", *Electrochimica Acta*, vol. 55, no. 28, pp. 8821–8828, 2010.

[KIT 11a] KITAURA H., HAYASHI A., TADANAGA K. *et al.*, "Improvement of electrochemical performance of all-solid-state lithium secondary batteries by surface modification of $LiMn_2O_4$ positive electrode", *Solid State Ionics*, vol. 192, no. 1, pp. 304–307, 2011.

[KIT 11b] KITAURA H., HAYASHI A., OHTOMO T. *et al.*, "Fabrication of electrode–electrolyte interfaces in all-solid-state rechargeable lithium batteries by using a supercooled liquid state of the glassy electrolytes", *Journal of Materials Chemistry*, vol. 21, no. 1, p. 118, 2011.

[KNU 83] KNUTZ B., SKAARUP S., "Cycling of $Li/Li_3N/TiS_2$ solid state cells", *Solid State Ionics*, vol. 9, pp. 371–374, 1983.

[KOB 08] KOBAYASHI T., IMADE Y., SHISHIHARA D. *et al.*, "All solid-state battery with sulfur electrode and thio-LISICON electrolyte", *Journal of Power Sources*, vol. 182, no. 2, pp. 621–625, 2008.

[KOM 01] KOMIYA R., HAYASHI A., MORIMOTO H. *et al.*, "Solid state lithium secondary batteries using an amorphous solid electrolyte in the system $(100-x)(0.6Li_2S\cdot 0.4SiS_2)\cdot xLi_4SiO_4$ obtained by mechanochemical synthesis", *Solid State Ionics*, vol. 140, no. 1, pp. 83–87, 2001.

[KON 92] KONDO S., TAKADA K., YAMAMURA Y., "New lithium ion conductors based on $Li_2S–SiS_2$ system", *Solid State Ionics*, vol. 53, pp. 1183–1186, 1992.

[KOT 12] KOTOBUKI M., "The current situation and problems of rechargeable lithium ion batteries", *The Open Electrochemistry Journal*, vol. 4, pp. 28–35, 2012.

[KUH 79] KUHS W.F., NITSCHE R., SCHEUNEMANN K., "The argyrodites—a new family of tetrahedrally close-packed structures", *Materials Research Bulletin*, vol. 14, no. 2, pp. 241–248, 1979.

[LAL 14] LALÈRE F., LERICHE J.-B., COURTY M. *et al.*, "An all-solid state NASICON sodium battery operating at 200 °C", *Journal of Power Sources*, vol. 247, pp. 975–980, 2014.

[LAL 15] LALÈRE F., Développement de batteries céramiques "tout-solide" Li- et Na-ion à haute température : synthèses et mécanismes électrochimiques, PhD Thesis, Université de Picardie Jules Verne, France, 2015.

[LAR 15] LARCHER D., TARASCON J.M., "Towards greener and more sustainable batteries for electrical energy storage", *Nature Chemistry*, vol. 7, pp, 19–29, 2015.

[LEV 09] LEVASSEUR A, PECQUENARD B., VINATIER P. *et al.*, "Microbatteries – Microsources d'énergie en couches minces", *Techniques de l'Ingénieur*, ARTICLE/RÉF : D3342 V2, 2009.

[LEY 12] LEY M.B., BOULINEAU S., JANOT R. *et al.*, "New li ion conductors and solid state hydrogen storage materials: $LiM(BH_4)_3Cl$, M= La, Gd", *The Journal of Physical Chemistry C*, vol. 116, no. 40, pp. 21267–21276, 2012.

[LIG 93] LIGHTFOOT P., METHA M.A., BRUCE P.G., "Crystal structure of the polymer electrolyte poly(ethylene oxide)$_3$: $LiCF_3SO_3$", *Science*, vol. 262, pp. 883–885, 1993.

[LOT 17] LOTSCH B.V., MAIER J., "Relevance of solid electrolytes for lithium-based batteries: a realistic view", Journal of Electroceramics, pp. 1–14, 2017.

[MAC 99] MACGLASHAN G.S., ANDREEV Y.G., BRUCE P.G., "Structure of the polymer electrolyte poly (ethylene oxide)$_6$: $LiAsF_6$", *Nature*, vol. 398, no. 6730, p. 792, 1999.

[MAC 04] MACHIDA N., SHIGEMATSU T., "An all-solid-state lithium battery with sulfur as positive electrode materials", *Chemistry Letters*, vol. 33, no. 4, pp. 376–377, 2004.

[MAE 09] MAEKAWA H., MATSUO M., TAKAMURA H. *et al.*, "Halide-stabilized $LiBH_4$, a room-temperature lithium fast-ion conductor", *Journal of the American Chemical Society*, vol. 131, no. 3, pp. 894–895, 2009.

[MAN 17] MANTHIRAM A., YU X., WANG S., "Lithium battery chemistries enabled by solid-state electrolytes", Nature Reviews Materials, vol. 2, p. 16103, 2017.

[MAS 95a] MASQUELIER C., KAGEYAMA H., TAKEUCHI T. *et al.*, "Chemistry and structure analysis in the $Li_{4+x}B_xSi_{1-x}O_4$ solid solution", *Journal of power sources*, vol. 54, no. 2, pp. 448–451, 1995.

[MAS 95b] MASQUELIER C., TABUCHI M., TAKEUCHI T. *et al.*, "Influence of the preparation process on the cation transport properties of $Li_{4+x}M_xSi_{1-x}O_4$ (M= B, Al) solid electrolytes", *Solid State Ionics*, vol. 79, pp. 98–105, 1995.

[MAS 11] MASQUELIER C., "Solid electrolytes: lithium ions on the fast track", *Nature Materials*, vol. 10, no. 9, p. 649, 2011.

[MAS 13] MASOUD E.M., KHAIRY M., MOUSA M.A., "Electrical properties of fast ion conducting silver based borate glasses: application in solid battery", *Journal of Alloys and Compounds*, vol. 569, pp. 150–155, 2013.

[MAT 09] MATSUO M., REMHOF A., MARTELLI P. *et al.*, "Complex hydrides with $(BH_4)^-$ and $(NH_2)^-$ anions as new lithium fast-ion conductors", *Journal of American Chemical Society*, vol. 131, pp. 16389–16391, 2009.

[MAT 10] MATSUO M., SATO T., MIURA Y. *et al.*, "Synthesis and lithium fast-ion conductivity of a new complex hydride $Li_3(NH_2)2_I$ with double-layered structure", *Chemistry of Materials*, vol. 22, no. 9, pp. 2702–2704, 2010.

[MCS 17] Mineral Commodity Summaries, U.S. Geological Survey, available at: https://minerals.usgs.gov/minerals/pubs/commodity/lithium/mcs-2017-lithi.pdf, consulted on 29/08/2017, January 2017.

[MEN 84] MENETRIER M., LEVASSEUR V., DELMAS C. *et al.*, "New secondary batteries for room temperature applications using a vitreous electrolyte", *Solid State Ionics*, vol. 14, no. 3, pp. 257–261, 1984.

[MEU 89] MEUNIER G., DORMOY R., LEVASSEUR A., "New positive-electrode materials for lithium thin film secondary batteries", *Materials Science and Engineering: B*, vol. 3, nos 1–2, pp. 19–23, 1989.

[MIN 00] MINAMI T., HAYASHI A., TATSUMISAGO M., "Preparation and characterization of lithium ion-conducting oxysulfide glasses", *Solid State Ionics*, vol. 136, pp. 1015–1023, 2000.

[MIN 08] MINAMI K., MIZUNO F., HAYASHI A. *et al.*, "Structure and properties of the $70Li_2S{\cdot}(30-x)P_2S_5{\cdot}xP_2O_5$ oxysulfide glasses and glass–ceramics", *Journal of Non-Crystalline Solids*, vol. 354, no. 2, pp. 370–373, 2008.

[MIN 09] MINAMI K., HAYASHI A., UJIIE S. *et al.*, "Structure and properties of $Li_2S–P_2S_5–P_2S_3$ glass and glass–ceramic electrolytes", *Journal of Power Sources*, vol. 189, no. 1, pp. 651–654, 2009.

[MIN 10] MINAMI K., HAYASHI A., TATSUMISAGO M., "Preparation and characterization of lithium ion conducting $Li_2S–P_2S_5–GeS_2$ glasses and glass-ceramics", *Journal of Non-Crystalline Solids*, vol. 356, no. 44, pp. 2666–2669, 2010.

[MIN 11] MINAMI K., HAYASHI A., TATSUMISAGO M., "Electrical and electrochemical properties of glass–ceramic electrolytes in the systems $Li_2S–P_2S_5–P_2S_3$ and $Li_2S–P_2S_5–P_2O_5$", *Solid State Ionics*, vol. 192, pp. 122–125, 2011.

[MIY 11] MIYAZAKI R., KARAHASHI T., KUMATANI N. *et al.*, "Room temperature lithium fast-ion conduction and phase relationship of LiI stabilized $LiBH_4$", *Solid State Ionics*, vol. 192, no. 1, pp. 143–147, 2011.

[MIZ 80] MIZUSHIMA K., JONES P.C., WISEMAN P.J. *et al.*, "Li_xCoO_2 ($0 < x < -1$): a new cathode material for batteries of high energy density", *Materials Research Bulletin*, vol. 15, no. 6, pp. 783–789, 1980.

[MIZ 02] MIZUNO, F., HAMA, S., HAYASHI, A. *et al.*, "All solid-state lithium secondary batteries using high lithium ion conducting $Li_2S–P_2S_5$ glass-ceramics", *Chemistry Letters*, vol. 31, no. 12, pp. 1244–1245, 2002.

[MIZ 03] MIZUNO F., HAYASHI A., TADANAGA K. *et al.*, "All-solid-state lithium secondary batteries using a layer-structured $LiNi_{0.5}Mn_{0.5}O_2$ cathode material", *Journal of Power Sources*, vol. 124, no. 1, pp. 170–173, 2003.

[MIZ 05a] MIZUNO F., HAYASHI A., TADANAGA K. *et al.*, "Effects of conductive additives in composite positive electrodes on charge–discharge behaviors of all-solid-state lithium secondary batteries", *Journal of the Electrochemical Society*, vol. 152, no. 8, pp. A1499–A1503, 2005.

[MIZ 05b] MIZUNO F., HAYASHI A., TADANAGA K. *et al.*, "Design of composite positive electrode in all-solid-state secondary batteries with $Li_2S-P_2S_5$ glass–ceramic electrolytes", *Journal of Power Sources*, vol. 146, nos 1–2, pp. 711–714, 2005.

[MIZ 06] MIZUNO F., HAYASHI A., TADANAGA K. *et al.*, "High lithium ion conducting glass-ceramics in the system $Li_2S–P_2S_5$", *Solid State Ionics*, vol. 177, no. 26, pp. 2721–2725, 2006.

[MO 11] MO Y., ONG S. P., CEDER G., "First principles study of the $Li_{10}GeP_2S_{12}$ lithium super ionic conductor material", *Chemistry of Materials*, vol. 24, no. 1, pp. 15–17, 2011.

[MOR 99] MORIMOTO H., YAMASHITA H., TATSUMISAGO M. *et al.*, "Mechanochemical synthesis of new amorphous materials of $60Li_2S{\cdot}40SiS_2$ with high lithium ion conductivity", *Journal of the American Ceramic Society*, vol. 82, no. 5, pp. 1352–1354, 1999.

[MOR 00] MORIMOTO H., YAMASHITA H., TATSUMISAGO M. *et al.*, "Mechanochemical synthesis of the high lithium ion conductive amorphous materials in the systems $Li_2S–SiS_2$ and $Li_2S–SiS_2–Li_4SiO_4$", *Journal of the Ceramic Society of Japan*, vol. 108, no. 1254, pp. 128–131, 2000.

[MOT 02] MOTO Y., KAMIYAMA T., "Structure of the thio-LISICON, Li_4GeS_4", *Solid State Ionics*, vol. 154, pp. 789–794, 2002.

[MUR 02a] MURAYAMA M., KANNO R., IRIE M. *et al.*, "Synthesis of new lithium ionic conductor thio-LISICON-lithium silicon sulfides system", *Journal of Solid State Chemistry*, vol. 168, no. 1, pp. 140–148, 2002.

[MUR 02b] MURAYAMA M., KANNO R., KAWAMOTO Y. *et al.*, "Structure of the thio-LISICON, Li4GeS4", *Solid State Ionics*, vol. 154, pp. 789–794, 2002.

[MUR 07a] MURUGAN R., THANGADURAI V., WEPPNER W., "Lithium ion conductivity of $Li_{5+x}Ba_xLa_{3-x}Ta_2O_{12}$ (x= 0–2) with garnet-related structure in dependence of the barium content", *Ionics*, vol. 13, no. 4, pp. 95–203, 2007.

[MUR 07b] MURUGAN R., WEPPNER W., SCHMID-BEURMANN P. *et al.*, "Structure and lithium ion conductivity of bismuth containing lithium garnets $Li_5La_3Bi_2O_{12}$ and $Li_6SrLa_2Bi_2O_{12}$", *Materials Science and Engineering: B*, vol. 143, no. 1, pp. 14–20, 2007.

[NAG 90] NAGAURA T., TOWAZA K., "Lithium ion rechargeable battery", in KOZAWA A. (ed) *Progress in Batteries and Solar Cells*, vol. 9, JEC Press Inc., Brunswick, 1990.

[NAG 09] NAGAO M., KITAURA H., HAYASHI A. *et al.*, "Characterization of all-solid-state lithium secondary batteries using $Cu_xMo_6S_{8-y}$ electrode and $Li_2S–P_2S_5$ solid electrolyte", *Journal of Power Sources*, vol. 189, no. 1, pp. 672–675, 2009.

[NAK 10] NAKAJIMA K., KATOH T., INDA Y. *et al.*, "Lithium-Ion Conducting Glass-Ceramics: Properties and Application in Lithium Metal Batteries", *Symposium on Energy Storage Beyond Lithium Ion; Materials Perspective*, Oak Ridge National Laboratory, OHARA INC, 8th October, 2010.

[NAG 11a] NAGAO M., HAYASHI A., TATSUMISAGO M., "All-solid-state lithium secondary batteries with high capacity using black phosphorus negative electrode", *Journal of Power Sources*, vol. 196, no. 16, pp. 6902–6905, 2011.

[NAG 11b] NAGAO M., HAYASHI A., TATSUMISAGO M., "Sulfur–carbon composite electrode for all-solid-state Li/S battery with $Li_2S–P_2S_5$ solid electrolyte", Electrochimica Acta, vol. 56, no. 17, pp. 6055–6059, 2011.

[NAG 12a] NAGAO M., HAYASHI A., TATSUMISAGO M., "Fabrication of favorable interface between sulfide solid electrolyte and Li metal electrode for bulk-type solid-state Li/S battery", *Electrochemistry Communications*, vol. 22, pp. 177–180, 2012.

[NAG 12b] NAGAO M., HAYASHI A., TATSUMISAGO M., "High-capacity Li_2S–nanocarbon composite electrode for all-solid-state rechargeable lithium batteries", *Journal of Materials Chemistry*, vol. 22, pp. 10015–10020, 2012.

[NAG 13] NAGAO M., KANNO R., "All-solid-state Li–sulfur batteries with mesoporous electrode and thio-LISICON solid electrolyte", *Journal of Power Sources*, vol. 222, pp. 237–242, 2013.

[NEU 99] NEUDECKER B.J., ZUHR R.A., BATES J.B., "Lithium silicon tin oxynitride (Li_ySiTON): high-performance anode in thin-film lithium-ion batteries for microelectronics", *Journal of Power Sources*, vol. 81, pp. 27–32, 1999.

[NEU 00] NEUDECKER B.J., DUDNEY N.J., BATES J.B., "Lithium-free thin-film battery with in situ plated Li anode", *Journal of the Electrochemical Society*, vol. 147, no. 2, pp. 517–523, 2000.

[NIS 09] NISHIO Y., KITAURA H., HAYASHI, A. *et al.*, "All-solid-state lithium secondary batteries using nanocomposites of NiS electrode/$Li_2S–P_2S_5$ electrolyte prepared via mechanochemical reaction", *Journal of Power Sources*, vol. 189, no. 1, pp. 629–632, 2009.

[OHT 06] OHTA N., TAKADA K., ZHANG L. *et al.*, "Enhancement of the high-rate capability of solid-state lithium batteries by nanoscale interfacial modification", *Advanced Materials*, vol. 18, no. 17, pp. 2226–2229, 2006.

[OHT 07] OHTA N., TAKADA K., SAKAGUCHI I. *et al.*, "$LiNbO_3$-coated $LiCoO_2$ as cathode material for all solid-state lithium secondary batteries", *Electrochemistry Communications*, vol. 9, no. 7, pp. 1486–1490, 2007.

[OHA 10] Ohara Corporation, "Lithium-Ion Conducting Glass-Ceramics (LICGC™)", available at: http://www.oharacorp.com/lic-gc.html, consulted on 27/09/15, 2010.

[OHT 12] OHTA S., KOBAYASHI T., SEKI J. *et al.*, "Electrochemical performance of an all-solid-state lithium ion battery with garnet-type oxide electrolyte", *Journal of Power Sources*, vol. 202, pp. 332–335, 2012.

[OHZ 95] OHZUKU T., UEDA A., YAMAMOTO N., "Zero-strain insertion material of Li [$Li_{1/3}Ti_{5/3}$]O_4 for rechargeable lithium cells", *Journal of the Electrochemical Society*, vol. 142, no. 5, pp. 1431–1435, 1995.

[ONG 13] ONG S.P., MO Y., RICHARDS W.D. *et al.*, "Phase stability, electrochemical stability and ionic conductivity of the $Li_{10\pm1}MP_2X_{12}$ (M= Ge, Si, Sn, Al or P, and X= O, S or Se) family of superionic conductors", *Energy & Environmental Science*, vol. 6, no. 1, pp. 148–156, 2013.

[PAL 12] PALOMARES V., SERRAS P., VILLALUENGA I. *et al.*, "Na-ion batteries, recent advances and present challenges to become low cost energy storage systems", *Energy & Environmental Science*, vol. 5, pp. 5884–5901, 2012.

[PAR 13] PARK H., YOON T., MUN J. *et al.*, "A comparative study on thermal stability of two solid electrolyte interphase (SEI) films on graphite negative electrode", *Journal of The Electrochemical Society*, vol. 160, no. 9, pp. A1539–A1543, 2013.

[PAT 39] PATTERSON A.L., "The Scherrer formula for X-ray particle size determination", *Physical Review*, vol. 56, no. 10, pp. 978, 1939.

[PHA 12] PHAN V.P., PECQUENARD B., LE CRAS F., "High-performance all-solid-state cells fabricated with silicon electrodes", *Advanced Functional Materials*, vol. 22, no. 12, pp. 2580–2584, 2012.

[POI 11] POIZOT P., DOLHEM F., "Clean energy new deal for a sustainable world: from non-CO_2 generating energy sources to greener electrochemical storage devices", *Energy & Environmental Science*, vol. 4, pp. 2003–2019, 2011.

[PRA 86] PRADEL A., RIBES M., "Electrical properties of lithium conductive silicon sulfide glasses prepared by twin roller quenching", *Solid State Ionics*, vol. 18, pp. 351–355, 1986.

[PRO 98] PROSINI P.P., PASSERINI S., VELLONE R. *et al.*, *Journal of Power Sources*, vol. 75, p. 73, 1998.

[Press notice] Press notice, 3M and BOLLORE, from December 7, 2012. available at: solutions.3mfrance.fr/3MContentRetrievalAPI/BlobServlet?lmd=1354876731000&local=fr_FR&assetType=MMM_Image&assetId=1319243303351&blobAttribute=ImageFile.

[QUA 11] QUARTARONE E., MUSTARELLI P., "Electrolytes for solid-state lithium rechargeable batteries: recent advances and perspectives", *Chemical Society Reviews*, vol. 40, no. 5, pp. 2525–2540, 2011.

[RAO 11] RAO R P., SHARMA N., PETERSON V.K. *et al.*, "Studies of lithium argyrodite solid electrolytes for all-solid-state batteries", *Physica Status Solidi (a)*, vol. 208, no. 8, pp. 1804–1807, 2011.

[RAO 12] RAO R.P., SHARMA N., PETERSON V.K. *et al.*, "Variation in structure and Li^+-ion migration in argyrodite-type Li_6PS_5X (X = Cl, Br, I) solid electrolytes", *Journal of Solid State Electrochemistry*, vol. 16, pp. 1807–1813, 2012.

[RAO 13] RAO R.P., SHARMA N., PETERSON V.K. *et al.*, "Formation and conductivity studies of lithium argyrodite solid electrolytes using in-situ neutron diffraction", *Solid State Ionics*, vol. 230, pp. 72–76, 2013.

[REN 16] REN21, Rapport sur le statut mondial des énergies renouvelables 2016, available at: http://www.ren21.net/wp-content/uploads/2016/06/GSR_2016_KeyFindings_FRENCH.pdf, consulted on 29/08/2017, 2016.

[RIB 80] RIBES M., BARRAU B., SOUQUET J.L., "Sulfide glasses: Glass forming region, structure and ionic conduction of glasses in Na_2S–XS_2-(X= Si; Ge), Na_2S–P_2S_5 and Li_2S–GeS_2 systems", *Journal of Non-Crystalline Solids*, vol. 38, pp. 271–276, 1980.

[RIB 82] RIBES M., CARETTE B., MAURIN M., "Verres conducteurs ioniques du système Li_2S–GeS_2–LiI. Leur utilisation dans des générateurs "tout solide" à anode de lithium", *Journal de Physique*, vol. 43, no. C9, pp. 403–406, 1982.

[RIC 15] RICHARDS W.D., MIARA L.J., WANG Y. *et al.*, "Interface stability in solid-state batteries", *Chemistry of Materials*, vol. 28, no. 1, pp. 266–273, 2015.

[ROB 97] ROBERTSON A.D., WEST A.R., RITCHIE A.G., "Review of crystalline lithium-ion conductors suitable for high temperature battery applications", *Solid State Ionics*, vol. 4, no. 1, pp. 1–11, 1997.

[ROG 00] ROGEZ J., KNAUTH P., GARNIER A. *et al.*, "Determination of the crystallization enthalpies of lithium ion conducting alumino–silicate glasses", *Journal of Non-Crystalline Solids*, vol. 262, no. 1, pp. 177–182, 2000.

[ROS 93] ROSSEN E., REIMERS J.N., DAHN J.R., "Synthesis and electrochemistry of spinel LT-$LiCoO_2$", *Solid State Ionics*, vol. 62, nos 1–2, pp. 53–60, 1993.

[ROT 12] ROTH E.P., ORENDORFF C.J., "How electrolytes influence battery safety", *Electrochemical Society Interface*, vol. 21, pp. 45–49, 2012.

[SAK 08] SAKUDA A., KITAURA H., HAYASHI A. *et al.*, "Improvement of high-rate performance of all-solid-state lithium secondary batteries using $LiCoO_2$ coated with Li_2O–SiO_2 glasses", *Electrochemical and Solid-State Letters*, vol. 11, no. 1, pp. A1–A3, 2008.

[SAK 09a] SAKUDA A., HAYASHI A., TATSUMISAGO M., "Interfacial observation between $LiCoO_2$ electrode and Li_2S–P_2S_5 solid electrolytes of all-solid-state lithium secondary batteries using transmission electron microscopy", *Chemistry of Materials*, vol. 22, no. 3, pp. 949–956, 2009.

[SAK 09b] SAKUDA A., KITAURA H., HAYASHI A. *et al.*, "Modification of interface between $LiCoO_2$ electrode and Li_2S–P_2S_5 solid electrolyte using Li_2O–SiO_2 glassy layers", *Journal of the Electrochemical Society*, vol. 156, no. 1, pp. A27–A32, 2009.

[SAK 09c] SAKUDA A., KITAURA H., HAYASHI A. *et al.*, "All-solid-state lithium secondary batteries with oxide-coated $LiCoO_2$ electrode and $Li_2S–P_2S_5$ electrolyte", *Journal of Power Sources*, vol. 189, pp. 527–530, 2009.

[SAK 10] SAKUDA A., HAYASHI A., OHTOMO T. *et al.*, "$LiCoO_2$ electrode particles coated with $Li_2S–P_2S_5$ solid electrolyte for all-solid-state batteries", *Electrochemical and Solid-State Letters*, vol. 13, no. 6, pp. A73–A75, 2010.

[SAK 11] SAKUDA A., HAYASHI A., OHTOMO T. *et al.*, "All-solid-state lithium secondary batteries using $LiCoO_2$ particles with pulsed laser deposition coatings of $Li_2S–P_2S_5$ solid electrolytes", *Journal of Power Sources*, vol. 196, pp. 6735–6741, 2011.

[SCA 93] SCARMINIO J., TALLEDO A., ANDERSSON A.A. *et al.*, "Stress and electrochromism induced by Li insertion in crystalline and amorphous V_2O_5 thin film electrodes", *Electrochimica Acta*, vol. 38, p. 1637-1642, 1993.

[SCR 95] SCROSATI B., "Battery technology: challenge of portable power", *Nature*, vol. 373, pp. 557–558, 1995.

[SEE 15] SEEO, "DryLyte Battery Advantages", available at : http://www.seeo.com/ drylyte-battery-advntgs/, consulted on 25/09/2015, 2015.

[SHI 12] SHIMIZU N., "Toyota Prototypes All-solid-state Battery With 5x Higher Output Density, Nikkei Technology", available at: http://techon.nikkeibp.co.jp/english/NEWS_EN/20120926/241911/, consulted on 25/09/2015, 26 September, 2012.

[SHI 14] SHIN B.R., NAM Y.J., OH D.Y. *et al.*, "Comparative study of TiS_2/LiIn all-solid-state lithium batteries using glass-ceramic Li_3PS_4 and $Li_{10}GeP_2S_{12}$ solid electrolytes", *Electrochimica Acta*, vol. 146, pp. 395–402, 2014.

[SMI 03] SMITH G., *Transactions of the American Institute of Mining and Metallurgical Engineers*, vol. 34, p. 467, 1903.

[SON 16] SONG S., DUONG H.M., KORSUNSKY A.M. *et al.*, "A Na^+ superionic conductor for room-temperature sodium batteries", *Scientific Reports*, vol. 6, Article number: 32330, 2016.

[STA 10] STADLER F., FIETZEK C., "Crystalline Halide Substituted Li-Argyrodites as Solid Electrolytes for Lithium Secondary Batteries", *ECS Transactions*, vol. 25, no. 36, pp. 177–183, 2010.

[SUN 17] SUN C., LIU J., GONG Y. *et al.*, "Recent advances in all-solid-state rechargeable lithium batteries", Nano Energy, vol. 33, pp. 363–386, 2017.

[SUT 12] SUTHANTHIRARAJ S.A., SARUMATHI R., "A new silver ion conducting $SbI_3–Ag_4P_2O_7$ nanocomposite solid electrolyte", *Applied Nanoscience*, vol. 3, no. 6, pp. 501–508, 2012.

[TAK 96] TAKADA K., AOTANI N., IWAMOTO K. *et al.*, "Solid state lithium battery with oxysulfide glass", *Solid State Ionics*, vol. 86, pp. 877–882, 1996.

[TAK 13] TAKAHASHI K., HATTORI K., YAMAZAKI T. *et al.*, "All-solid-state lithium battery with $LiBH_4$ solid electrolyte", *Journal of Power Sources*, vol. 226, p. 61, 2013.

[TAN 14] TANIBATA N., NOI K., HAYASHI A. *et al.*, "Preparation and characterization of highly sodium ion conducting Na_3PS_4–Na_4SiS_4 solid electrolytes", *RSC Advances*, vol. 4, no. 33, pp. 17120–17123, 2014.

[TAN 15] TANIBATA N., MATSUYAMA T., HAYASHI A. *et al.*, "All-solid-state sodium batteries using amorphous TiS_3 electrode with high capacity", *Journal of Power Sources*, vol. 275, pp. 284–287, 2015.

[TAR 01] TARASCON J.M., ARMAND M., "Issues and challenges facing rechargeable lithium batteries", *Nature*, vol. 414, no. 6861, p. 359, 2001.

[TAR 15] TARHOUCHI I., Etude des phases $Li_{10}MP_2S_{12}$ (M=Sn, Si) comme électrolyte pour batteries tout-solide massives, PhD Thesis, Université de Bordeaux, France, 2015.

[TAT 97] TATSUMISAGO M., HIRAI K., MINAMI T. *et al.*, "Preparation and characterisation of superionic Li_2S–SiS_2–Li_4GeO_4 glasses", *Physics and Chemistry of Glasses*, vol. 38, no. 2, pp. 63–65, 1997.

[TAT 00] TATSUMISAGO M., YAMASHITA H., HAYASHI A. *et al.*, "Preparation and structure of amorphous solid electrolytes based on lithium sulfide", *Journal of Non-crystalline Solids*, vol. 274, no. 1, pp. 30–38, 2000.

[TAT 02] TATSUMISAGO M., HAMA S., HAYASHI A. *et al.*, "New lithium ion conducting glass-ceramics prepared from mechanochemical Li_2S–P_2S_5 glasses", *Solid State Ionics*, vol. 154, pp. 635–640, 2002.

[TAT 08] TATSUMISAGO M., HAYASHI A., "All-solid-state lithium secondary, batteries using sulfide-based glass ceramic electrolytes", *Functional Materials Letters*, vol. 1, pp. 31–36, 2008.

[TAT 13] TATSUMISAGO M., NAGAO M., HAYASHI A., "Recent development of sulfide solid electrolytes and interfacial modification for all-solid-state rechargeable lithium batteries", *Journal of Asian Ceramic Societies*, vol. 1, pp. 17–25, 2013.

[THA 03] THANGADURAI V., KAACK H., WEPPNER W.J., "Novel fast lithium ion conduction in Garnet-type $Li_5La_3M_2O_{12}$ (M= Nb, Ta)", *Journal of the American Ceramic Society*, vol. 86, no. 3, pp. 437–440, 2003.

[THA 05a] THANGADURAI V., WEPPNER W., "$Li_6ALa_2Nb_2O_{12}$ (A= Ca, Sr, Ba): a new class of fast lithium ion conductors with garnet-like structure", *Journal of the American Ceramic Society*, vol. 88, no. 2, pp. 411–418, 2005.

[THA 05b] THANGADURAI V., SCHWENZEL J., WEPPNER W., "Tailoring ceramics for specific applications: a case study of the development of all-solid-state lithium batteries", *Ionics*, vol. 11, nos 1–2, pp. 11–23, 2005.

[THA 05c] THANGADURAI V., SCHWENZEL J., WEPPNER W., "Tailoring ceramics for specific applications: a case study of the development of all-solid-state lithium batteries", Conference: Patras Conference on Solid State Ionics – Transport Properties, *Ionics*, vol. 11, nos 1–2, pp. 11–23, 2005.

[THA 06a] THANGADURAI V., WEPPNER W., "Effect of sintering on the ionic conductivity of garnet-related structure $Li_5La_3Nb_2O_{12}$ and In-and K-doped $Li_5La_3Nb_2O_{12}$", *Journal of Solid State Chemistry*, vol. 179, no. 4, pp. 974–984, 2006.

[THA 06b] THANGADURAI V., WEPPNER W., "Recent progress in solid oxide and lithium ion conducting electrolytes research", *Ionics*, vol. 12, no. 1, pp. 81–92, 2006.

[TOY 95-18] Toyota Global, "Research on next-generation batteries that surpass lithium-ion batteries", available at : http://www.toyota-global.com/innovation/environmental_ technology/keytech/, consulted on 01/13/2018, 1995-2018.

[TUB 14] TUBANDT C., LORENZ E., "Molecular state and electrical conductivity of crystallised salts", *Zeitschrift für physikalische Chemie, Stöchiometrie und Verwandtschaftslehre*, vol. 87, no. 5, pp. 513–542, 1914.

[UJI 13] UJIIE S., HAYASHI A., TATSUMISAGO M., "Preparation and ionic conductivity of $(100-x)(0.8Li_2S{\cdot}0.2P_2S_5){\cdot}x$LiI glass–ceramic electrolytes", *Journal of Solid State Electrochemistry*, vol. 17, no. 3, pp. 675–680, 2013.

[UNE 15] UNEMOTO A., IKESHOJI T., YASAKU S. *et al.*, "Stable interface formation between TiS_2 and $LiBH_4$ in bulk-type all-solid-state lithium batteries", *Chemistry of Materials*, vol. 27, pp. 5407–5416, 2015.

[VAN 07] VAN WÜLLEN L., ECHELMEYER T., MEYER H.W. *et al.*, "The mechanism of Li-ion transport in the garnet $Li_5La_3Nb_2O_{12}$", *Physical Chemistry Chemical Physics*, vol. 9, no. 25, pp. 3298–3303, 2007.

[VEN 98] VENKATESWARLU M., SATYANARAYANA N., "AC conductivity studies of silver based fast ion conducting glassy materials for solid state batteries", *Materials Science and Engineering: B*, vol. 54, no. 3, pp. 189–195, 1998.

[VIN 93] VINCENT C.A., SCROSATI B., *Modern Batteries. An Introduction to Electrochemical Power Sources*, 2nd ed., Arnold, London, 1993.

[VIZ 14] VIZINTIN A., PATEL M.U.M., GENORIO B. *et al.*, "Effective Separation of Lithium Anode and Sulfur Cathode in Lithium–Sulfur Batteries", ChemElectroChem, vol. 1, pp. 1040–1045, 2014.

[WAD 83] WADA H., MENETRIER M., LEVASSEUR A. *et al.*, "Preparation and ionic conductivity of new B_2S_3–Li_2S–LiI glasses", *Materials Research Bulletin*, vol. 18, no. 2, pp. 189–193, 1983.

[WAN 96] WANG B., BATES J.B., HART F.X. *et al.*, "Characterization of thin-film rechargeable lithium batteries with lithium cobalt oxide cathodes", *Journal of the Electrochemical Society*, vol. 143, no. 10, pp. 3203–3213, 1996.

[WEI 86] WEISBACH A., *Neues Jahrbuch für Mineralogie, Geologie und Paläontologie*, vol. 2, p. 67, 1886.

[WEN 15] WENZEL S., LEICHTWEISS T., KRÜGER D. *et al.*, "Interphase formation on lithium solid electrolytes—an in situ approach to study interfacial reactions by photoelectron spectroscopy", *Solid State Ionics*, vol. 278, pp. 98–105, 2015.

[WEO 12] WEO-2012 (World Energy Outlook 2012), International Energy Agency, visited 29/08/2017.

[WHI 04] WHITTINGHAM M.S., "Lithium batteries and cathode materials", *Chemical Reviews*, vol. 104, no. 10, pp 4271–4302, 2004.

[WIE 89] WIECZOREK W., SUCH K., WYCISLIK H. *et al.*, "Modifications of crystalline structure of peo polymer electrolytes with ceramic additives", *Solid State Ionics*, vol. 36, pp. 225–227, 1989.

[WRI 75] WRIGHT P.V., "Electrical conductivity in ionic complexes of poly(ethylene oxide), *Polymer International*, vol. 7, no. 5, pp. 319–327, 1975.

[WU 03] WU Y.P., RAHM E., HOLZE R., "Carbon anode materials for lithium ion batteries", *Journal of Power Sources*, vol. 114, no. 2, pp. 228–236, 2003.

[XIE 11] XIE H., GOODENOUGH J.B., LI Y., "$Li_{1.2}Zr_{1.9}Ca_{0.1}(PO_4)_3$, a room-temperature Li-ion solid electrolyte", *Journal of Power Sources*, vol. 196, no. 18, pp. 7760–7762, 2011.

[YAK 97] YAKUBOVICH O.V., URUSOV V.S., "Electron density distribution in lithiophosphatite Li_3PO_4; crystallochemical features of orthophosphates with hexagonal close packing", *Crystallography Reports*, vol. 42, no. 2, pp. 261–268, 1997.

[YAM 07] YAMANE H., SHIBATA M., SHIMANE Y. *et al.*, "Crystal structure of a superionic conductor, $Li_7P_3S_{11}$", *Solid State Ionics*, vol. 178, no. 15, pp. 1163–1167, 2007.

[YAN 03] YANG S., SONG Y., NGALA K. *et al.*, "Performance of $LiFePO_4$ as lithium battery cathode and comparison with manganese and vanadium oxides", *Journal of Power Sources*, vol. 119, pp. 239–246, 2003.

[YAN 12] YANG Y., ZHENG G., MISRA S. *et al.*, "High-capacity micrometer-sized Li_2S particles as cathode materials for advanced rechargeable lithium-ion batteries", *Journal of the American Chemical Society*, vol. 134, no. 37, pp. 15387–15394, 2012.

[YAO 67] YAO Y.F.Y., KUMMER J.T., "Ion exchange properties and rates of ionic diffusion in beta-alumina", *Journal of Inorganic & Nuclear Chemistry*, vol. 29, no. 9, p. 2453, 1967.

[YAO 08] YAO W., BERG K., MARTIN S., "Structure and properties of glasses in the MI + M_2S + ($0.1Ga_2S_3$ + $0.9GeS_2$), M= Li, Na, K and Cs, system", *Journal of Non-Crystalline Solids*, vol. 354, no. 18, pp. 2045–2053, 2008.

[YAO 16] YAO X., HUANG B., YIN J. *et al.*, "All-solid-state lithium batteries with inorganic solid electrolytes: review of fundamental science", *Chinese Physics B*, vol. 25, no. 1, 018802, pp. 1–14, 2016.

[YIN 15] YIN J., YAO X., PENG G. *et al.*, "Influence of the Li–Ge–P–S based solid electrolytes on NCA electrochemical performances in all-solid-state lithium batteries", *Solid State Ionics*, vol. 274, pp. 8–11, 2015.

[YOO 13] YOON Y., PARK C., KIM J. *et al.*, "Lattice orientation control of lithium cobalt oxide cathode film for all-solid-state thin film batteries", *Journal of Power Sources*, vol. 226, pp. 186–190, 2013.

[YU 97] YU X., BATES J.B., JELLISON G.E. *et al.*, "A stable thin-film lithium electrolyte: lithium phosphorus oxynitride", *Journal of the Electrochemical Society*, vol. 144, no. 2, pp. 524–532, 1997.

[YUB 15] YUBUCHI S., HAYASHI A., TATSUMISAGO M., "Sodium-ion conducting Na_3PS_4 electrolyte synthesized via a liquid-phase process using N-methylformamide", *Chemistry Letters*, vol. 44, no. 7, pp. 884–886, 2015.

[ZAI 13] ZAINOL N.H., SAMIN S.M., OTHMAN L. *et al.*, "Magnesium ion-based gel Polymer Electrolytes: ionic conduction and infrared spectroscopy studies", *International Journal of Electrochemical Science*, vol. 8, no. 3, pp. 3602–3614, 2013.

[ZHA 90] ZHANG Z., KENNEDY J.H., "Synthesis and characterization of the B_2S_3–Li_2S, the P_2S_5–Li_2S and the B_2S_3–P_2S_5–Li_2S glass systems", *Solid State Ionics*, vol. 38, nos 3–4, pp. 217–224, 1990.

[ZHA 12] ZHAO Y., DAEMEN L.L., "Superionic conductivity in lithium-rich anti-perovskites", *Journal of the American Chemical Society*, vol. 134, no. 36, pp. 15042–15047, 2012.

[ZHA 14] ZHANG J., HAN J., ZHU J. *et al.*,"High pressure-high temperature synthesis of lithium-rich $Li_3O(Cl, Br)$ and $Li_{3-x}Ca_{x/2}OCl$ anti-perovskite halides", *Inorganic Chemistry Communications*, vol. 48, pp. 140–143, 2014.

[ZHA 17] ZHAN L., ZHANG Y., ZHUANG X. *et al.*, "Ionic conductivities of lithium borohydride-lithium nitride composites", *Solid State Ionics*, vol. 304, pp. 150–155, 2017.

Index

F, G

H, I

L

M , N

P, R

S

T, V

Other titles from

ISTE

in

Energy

2017

BROSETA Daniel, RUFFINE Livio, DESMEDT Arnaud
Gas Hydrates 1: Fundamentals, Characterization and Modeling

LA SCALA Massimo
From Smart Grids to Smart Cities: New Challenges in Optimizing Energy Grids

MOLINA Géraldine, MUSY Marjorie, LEFRANC Margot
Building Professionals Facing the Energy Efficiency Challenge

SIMON Patrice, BROUSSE Thierry, FAVIER Frédéric
Supercapacitors Based on Carbon or Pseudocapacitive Materials (Energy Storage – Batteries, Supercapacitors Set – Volume 3)

2016

ALLARD Bruno
Power Systems-on-Chip: Practical Aspects of Design

ANDRE Michel, SAMARAS Zissis
Energy and Environment

DUFOUR Anthony
Thermochemical Conversion of Biomass for the Production of Energy and Chemicals

2015

CROGUENNEC Laurence, MONCONDUIT Laure, DEDRYVÈRE Rémi
Electrodes for Li-ion Batteries
(Energy Storage – Batteries, Supercapacitors Set – Volume 2)

LEPRINCE-WANG Yamin
Piezoelectric ZnO Nanostructure for Energy Harvesting
(Nanotechnologies for Energy Recovery Set – Volume 1)

ROBYNS Benoît, FRANÇOIS Bruno, DELILLE Gauthier, SAUDEMONT Christophe
Energy Storage in Electric Power Grids

ROSSI Carole
Al-based Energetic Nanomaterials
(Nanotechnologies for Energy Recovery Set – Volume 2)

TARASCON Jean-Marie, SIMON Patrice
Electrochemical Energy Storage
(Energy Storage – Batteries, Supercapacitors Set – Volume 1)

2013

LALOUI Lyesse and DI DONNA Alice
Energy Geostructures: Innovation in Underground Engineering

2012

BECKERS Benoit
Solar Energy at Urban Scale

ROBYNS Benoît, DAVIGNY Arnaud, FRANÇOIS Bruno, HENNETON Antoine, SPROOTEN Jonathan
Electricity Production from Renewable Energies

2011

GAO Fei, BLUNIER Benjamin, MIRAOUI Abdellatif
Proton Exchange Membrane Fuel Cell Modeling

MULTON Bernard
Marine Renewable Energy Handbook

2010

BRUNET Yves
Energy Storage

2009

SABONNADIÈRE Jean-Claude
Low Emission Power Generation Technologies and Energy Management

SABONNADIÈRE Jean-Claude
Renewable Energy Technologies

Printed and bound by CPI Group (UK) Ltd, Croydon, CR0 4YY